METEORITES

Schiffer Publishing Ltd

4880 Lower Valley Road, Atglen, Pennsylvania 19310

Bruce L. Stinchcomb

Acknowledgments

Meteorites can be a wonderful obsession! The author wishes to thank the following obsessed(?) persons for contributions, either intellectual or physical: Chris Baught; Charlie Burkhardt of Florissant Valley Community College; Richard Hagar of Imperial, Missouri; the late Edward Friton; John Schooler of Schooler's Minerals, Fossils, and Meteorites; and other persons who may have been overlooked in the quixotic search for meteorites and their meaning.

Other Schiffer Books By The Author:
Cenozoic Fossils I: Paleogene. ISBN: 9780764334245. $29.99
Cenozoic Fossils II: The Neogene. ISBN: 9780764335808. $29.99
Mesozoic Fossils I: Triassic and Jurassic. ISBN: 9780764331633. $29.99
Mesozoic Fossils II: The Cretaceous Period. ISBN: 9780764332593. $29.99
Paleozoic Fossils. ISBN: 9780764329173. $29.95
World's Oldest Fossils. ISBN: 9780764326974. $29.95

Other Schiffer Books on Related Subjects:
Maryland's Geology. Martin F. Schmidt, Jr. ISBN: 9780764335938. $24.99

Copyright © 2011 by Bruce L. Stinchcomb

Library of Congress Control Number: 2011924874

Designed by Mark David Bowyer
Type set in Bodoni Bd BT / Humanist 521 BT

ISBN: 978-0-7643-3728-4
Printed in China

Schiffer Books are available at special discounts for bulk purchases for sales promotions or premiums. Special editions, including personalized covers, corporate imprints, and excerpts can be created in large quantities for special needs. For more information contact the publisher:

Published by Schiffer Publishing Ltd.
4880 Lower Valley Road
Atglen, PA 19310
Phone: (610) 593-1777; Fax: (610) 593-2002
E-mail: Info@schifferbooks.com

For the largest selection of fine reference books on this and related subjects, please visit our website at
www.schifferbooks.com
We are always looking for people to write books on new and related subjects. If you have an idea for a book please contact us at the above address.

This book may be purchased from the publisher.
Include $5.00 for shipping.
Please try your bookstore first.
You may write for a free catalog.

In Europe, Schiffer books are distributed by
Bushwood Books
6 Marksbury Ave.
Kew Gardens
Surrey TW9 4JF England
Phone: 44 (0) 20 8392 8585; Fax: 44 (0) 20 8392 9876
E-mail: info@bushwoodbooks.co.uk
Website: www.bushwoodbooks.co.uk

Contents

Introduction

Looking up into a clear, light-pollution-free sky with its thousands of stars can be an awesome experience—a glimpse or insight into the immenseness of space. When accompanied with the fact that some of the objects being observed, like the Andromeda Galaxy, are millions of light years away, the awesomeness of the experience is augmented. This sensing, this awareness, of both immense distance and time, the two faces of the space-time continuum, is one of the great appeals of astronomy. Visually experiencing the electromagnetic emissions of energy (light) of our part of the Milky Way Galaxy is a phenomena associated with energy—but, is there a counterpart to this phenomena involving matter (or mass), the other side of the mass-energy "equation?" The answer is yes, there is—its in meteorites—those samples of matter that periodically come from space to Earth and, like the energy of star light, can also be "messengers" from distant worlds of both space and time.

Among other things meteorites are—

- The oldest and most ancient "things" which one can actually touch—some of their components having been created **before** the formation of the Solar System.

- The most primitive forms of matter which you can actually see and handle—the assemblage of the matter in some of them originating from stars that no longer exist—stars which, in their "death throes", threw off matter forming nebulae, masses of dust, and gas that became the birthplace of later stars.

- Representatives of that part of cosmic time that exceeds the age of any planets of the Solar System—some meteorites being samples of the original material from which the planets and the Sun were made.

The rocks shown here are really different as a collectible! Whereas conventional minerals and rocks record such phenomena as ancient life (fossils), chemical activity coming from below the Earth's surface as volcanoes and related phenomena (minerals); the collectible rocks shown here tell of quite different phenomena—phenomena such as nebula, nova and supernova, red giant stars, and impacting celestial bodies. These are phenomena associated with astronomy, not geology! Meteorites, these rocks from space, really are "out of this world" and remain one of the most exotic and interesting (as well as information-rich) collectible one can get into.

Chapter 1

A First Order Look at the Only Collectible from "Other Worlds"

Wonderful Rocks from Space!

Meteorites represent the only tangible objects (other than moon rocks) which a person actually can touch and own (other space rocks like lunar samples, obtained by the U. S. and Soviet space programs, are generally locked up and curated, only infrequently are they placed on exhibit). During recent decades, a surprising number of meteorites have surfaced worldwide and some of these, often the majority, have come onto the geo-collectible market. This "flood" of meteorites has resulted in an unprecedented opportunity for persons to acquire, and become familiar with, on a first hand basis, a form of matter that actually comes from space.

Meteorites come in three basic types: **metal *(siderites)*, stony (stones),** and ***stony-iron***—the latter being those meteorites which are intermediate between metallic and stony types.

Phenomena Associated with Space Rocks (Meteoroids) Entering the Earth's Atmosphere

Rocks in space are known as meteoroids. Meteoroids usually occupy various orbits around the Sun, their orbits sometimes coinciding with that of the Earth. When this happens, the **meteoroid** may be on its way to becoming a **meteorite**. A meteoroid upon entering the earth's atmosphere is slowed down from its cosmic velocity by friction. This heats up the meteoroid (which then may become a meteor). Atmospheric entry then heats its surface where extreme friction is generated as it plunges into the atmosphere and in the process produces a plasma with accompanying spectacular light and aerial effects. This intense heating causes surface material of the meteor to vaporize or to ablate. The ablation process will be preceded by melting, which will produce what is known as a fusion crust. The ablation process also effectively slows down the meteor (which is what the tiles on the Space Shuttle also were for) allowing it to impact or it may be totally destroyed by ablation, depending upon how large it originally was. Its composition also plays a part in its survival, as does its angle and speed of entry into the earth's atmosphere. The process of ablation leaves a **distinctive signature** on the surface of the meteor which, if it lands, now becomes a **meteorite**. Later weathering of the meteorite on the surface of the earth may modify its ablation surface to produce an oxide crust or weathering may modify the entire meteorite, producing what is known as terrestrialization (or terrestrization). Meteorites have been entering the earth's environment for millions (or billions) of years. Those which are actually seen to arrive on the earth's surface are known as **falls**. Those which fell sometime in the past (sometime even in the geologic past) and found at a later time are known as **finds**.

Meteorites and Glacial Erratics

Over a large portion of the earth's northern hemisphere rocks are found of various types which were carried hundreds of miles by the movement of continental glaciers during the earth's geologically recent Ice Age. These transported rocks often are of types not native to where they were found and are referred to as glacial *erratics*. Glacial erratics represent fragments of outcrops transported from (sometimes) distant locations—they can provide information on the geology of areas hundreds of miles away from where they were found. In a similar way meteorites can be considered as transported erratics—but in this case transported through space from other worlds—viz. celestial erratics. Celestial or extraterrestrial erratics which, like those of the glacial kind, are fragments of outcrops—in this case pieces of asteroids and/or planets transported through space from many distant locations and, through the processes and "kindness" of nature, brought to the Earth.

Side view of same specimen showing ablation surface at the side of the meteorite.

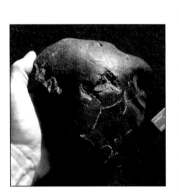

Ablation surface on a **stony** meteorite. Ablation is the process by which a meteor's surface material is removed by friction as the meteor enters the earth's atmosphere. It occurs from the intense friction generated during the meteor's travel through the atmosphere. Note pits (remaglyphs) at the right; these were produced by eddies which formed on the meteor's surface during the ablation process. A few pieces at the tip of this "nose cone" have been removed by weathering. Before entering the earth's atmosphere, this space rock was considerably larger than it is now—the material removed having been lost from ablation. Find-NWA (northwest Africa). (Value range C).

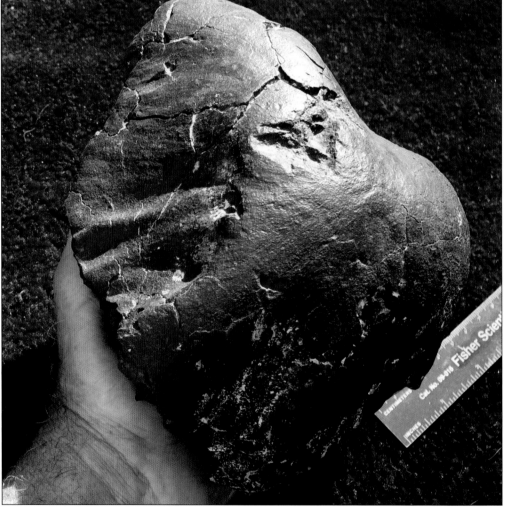

"Nose cone" and troughs from ablation on previously shown meteorite specimen.

Fusion crust on a recent fall. Note burnt look and darkening. Also notice small bubbles in this dark layer. Such bubbles (or small vesicles) generally are not found with meteorites unless they are associated with fusion crust. Fusion crusts generally are relatively thin, like this, for as soon as the fusion crust forms, it usually is removed by ablation. Sometimes the rate of ablation exceeds the rate of thermal conductivity so that a meteorite collected just after falling can be cold.

Close-up view of fresh fusion crust.

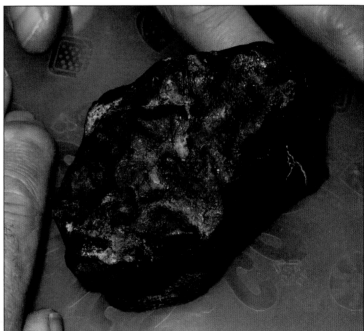

Weathered ablation surface of northwest Africa (NWA) meteorite with shallow remaglypts of a type characteristic of stony meteorites.

Oxide crust and ablation surface showing remaglypts. Remaglypts, thumb-print-like depressions, are generated by ablation and are especially characteristic of metallic meteorites—stony meteorites often exhibit ablation surfaces like those of the previous photos. The surface of this meteorite has been more affected by terrestrial weathering than were the previously shown specimens.

Smaller stony NWA with a somewhat weathered fusion crust showing cracks enlarged from terrestrial (atmospheric) weathering since its entrance into the earth's environment. The crack may have already been present before entry, perhaps produced from an encounter with another meteor when in orbit around the Sun.

Buffed, large stony NWA on stand. Do the numerous small pits (remaglypts) on this large stone resemble fish scales of a "tasty fish" to this kitty? (Value range A).

Large, mounted NWA stony meteorite with a flat surface which exhibits shallow remaglypts. This plane surface originated as a crack in the parent asteroid. It may have been produced from asteroidal impact or originated from forces caused by differential cooling between the center of the asteroid and its outer shell. Similar cracks on earth rocks can develop from forces responsible for earth movements (tectonic forces)—these are known as joints. (Value range A).

Stony meteorite with smooth ablation surface and some terrestrial weathering as well as crack formation, (NWA specimen).

Weathered fusion crust with broad, shallow remaglypts. (Stony NWA).

Geologically recent impact site on Earth. Meteor Crater Arizona (Also known as the Barringer or Winslow Crater) is one of the most spectacular and best preserved impact sites on the Earth's surface. Because of the Earth's atmosphere, terrestrial impact sites like this generally are destroyed by erosion and weathering after a few million years. The impact's effect on underlying rocks, however, may still be preserved and observable—such an impact structure, where the impact crater itself has been removed but the effects of impact can still be seen in underlying rock strata, **is known as an astroblem.**

Impact Phenomena—
Shatter Cones

High velocity impact of a meteor with the earth or another planet or impact between two meteors can produce what are known as shatter cones. Shatter cones represent a distinct phenomena associated with high velocity impact. They are more or less cone shaped, have a distinct surface pattern and are associated with highly broken rock known as breccia.

Back side of the above specimen showing striations characteristic of shatter cones.

Monomict breccia associated with shatter cones of the Steinheim Basin. Such breccia is produced from high velocity impact and is generally associated with shatter cones. The term monomict, a term used in meteoritics, refers to the fact that all of the clasts in a breccia are composed of the **same material** (in contrast to a polymict breccia in which the clasts come from different sources and therefore can consist of different kinds of clasts).

Shatter cones, Steinheim Basin, Germany. These are particularly nice shatter cones. Shatter cones are produced from high velocity impact when a meteor hits either a planetary surface or another meteor. The radiating striations are particularly characteristic of shatter cones—they possess a distinctive signature, which similar looking objects such as "cone-in-cone structures" lack. (Value range E).

A small, well formed shatter cone, Crooked Creek Structure, southern Missouri. The radiating striations shown here are typical of shatter cones. (Value range F).

Striations can be seen on these shatter cones from the Crooked Creek Structure, but they are less pronounced than are those on the cones from the Steinheim Basin.

A group of small shatter cones from the Crooked Creek Astroblem, Steelville, Missouri.

Another shatter cone from the Crooked Creek Astroblem.

Impact Phenomena, Breccia, and Brecciation

Breccia is a rock composed of angular rock fragments. With meteorites, and also in what are known as impact breccias, the angular fragments were produced from the energy of impact. **Brecciation** is the process of forming a breccia. With meteorites, brecciation is produced by the shattering of impacting materials. Brecciation is characterized by the existence of **sharp**, **angular fragments** known as **clasts**. *Breccias can be produced in various ways, but one of the most effective ways to produce a breccia and brecciation is with high velocity impact.*

Breccia, Decaturville astroblem, Missouri. This breccia came from the Decaturville Astroblem, which is of impact origin as both shatter cones and shocked quartz are associated with it. Note that the clasts are well separated from each other—a characteristic of impact breccias collected near "ground zero" (the center of the structure) where this specimen was collected. (Value range G).

Monomict breccia. Clasts of chert (sedimentary quartz), in this breccia from southern Missouri, have been cemented with manganese oxide (psilomelane). Although found clustered together in an area where satellite imagery shows a circular structure—a circular structure which might indicate an asteroid impact—this breccia is probably not of impact origin. The process of producing a breccia like this is known as brecciation. Many phenomena besides impact can produce breccias, some of which include earth movement (faulting), subsurface weathering, and volcanic activity. This breccia probably originated from some sort of subsurface weathering phenomena; a process which can also deposit manganese minerals, the black mineral that occurs between the chert clasts. (Value range G).

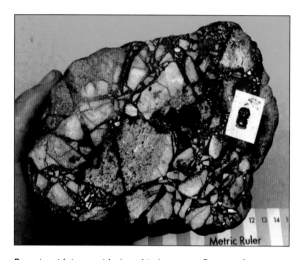

Breccia with iron oxide (goethite) cement. Breccias from an astroblem like the Crooked Creek structure can be cemented with iron oxide—iron possibly derived from material of the parent asteroid responsible for the structure. Unless nickel is associated with such iron minerals however, there is no proof of the iron's extra-terrestrial source. (Value range G).

Breccia. Clasts in these breccias are composed of chert, which is silica (quartz), a common chemically precipitated sedimentary rock found on the Earth. The matrix between the clasts is rock flour—in this case highly pulverized chert. Shatter cones have **not** been found associated with these breccias, **so their extraterrestrial origin is questioned,** but rock flour is generated in quantity with high velocity impact between both asteroids and the Earth's surface. **Shatter cones** and **shocked quartz** are two criteria which definitively can determine a structure's extraterrestrial origin. Douglas County knobs, southern Missouri. (Value range G).

Breccia, Vredefort Dome, South Africa. This breccia came from a very large and geologically ancient astroblem in southern Africa. Note that it is a **polymict breccia**—its clasts consist of materials **derived from multiple sources.** Note also that the clasts are well separated from each other—a characteristic of impact breccias collected from near "ground zero." This scattered-clast breccia forms from extremely violent, high-velocity impact phenomena characterized by lots of energy. (Value range G).

Breccia with ferruginous (iron bearing) cement. Compare this specimen with one at the bottom of this page. In this and the bottom specimen all geologic variables except one remain constant—the one which differs is the condition of the clasts. Note here that the clasts are angular—a characteristic of a breccia. Compare these clasts with those of the bottom photo.

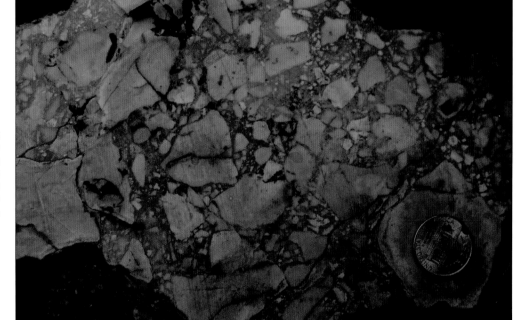

Angular clasts in a peculiar breccia associated with the Douglas County Knobs (Southern Missouri). See Chapter 8. Material filling space between the clasts appears to be rock flour—finely pulverized material also produced by high velocity impact.

This is a conglomerate and **not** a breccia! Note the shape of the clasts! Conglomerates differ from breccias in the **shape** of the clasts. Clasts of a conglomerate are **rounded**; those of a breccia are angular and are **not** rounded or abraded. Conglomerates are typical (and common) terrestrial rocks. The rounding of the clasts was a consequence of their being carried by and abraded in a stream sometime during the geologic past. Streams existing for a long period of time and capable of abrading and rounding rocks are **characteristic of the Earth** but are not found in any other place in the Solar System. (The streams on Mars probably carried some rocks but not long enough or far enough to extensively abrade and round them as happens on Earth). **The clasts found in meteorites never exhibit such rounding**.

Brecciation in a weathered NWA meteorite. In this meteorite, the clasts of the breccia have **not** been totally separated from each other. Impact which produced the breccia involved a collision between two objects in space before the meteor entered the earth. Breccias produced by more violent impact will have their clasts totally separated from each other and clasts of different rock types will often be mixed together (polymict breccia). Shown here is a **monomict** breccia—that is all of the clasts came from the same region and are compositionally the same. Note the prominent crack dividing the specimen. (Value range E).

Stony meteorite containing a dark colored clast. A large, prominent clast (dark colored mass) lies in the middle of this NWA-L4 chondrite. Another smaller clast is at 2:00 o'clock smaller from the prominent one. (Value range E).

Polymict breccia. Note the large bluish clast at 10:00 (at upper left) in this NWA (northwest Africa) meteorite. Note also the dark, homogeneous clast (carbonaceous chondrite clast?) at 6:00 (at the bottom).

Chondrules

A phenomena found in many (most?) stony meteorites are small (2-8 mm in diameter) spheres which to the experienced meteorist, because of their distinctive signature, are considered unique to meteorites. **Chondrules are a cosmic phenomena,** the product of poorly understood mechanisms associated with nebular evolution and associated with matter ejected from stars during their last stages of existence. These interesting and puzzling astronomical their objects will be discussed in later portions of this book.

Metal in Meteorites

Meteorites usually contain some metal. Generally, this metal is an alloy of nickel and iron. This is true even for most stony meteorites, whereas metallic meteorites are almost **entirely made of metal. The metallic component of meteorites is a consequence of the primitive condition of the matter of which they are made**—especially when compared to earth rocks, which normally don't contain elemental metal. With earth rocks any metallic nickel-iron that existed in the matter from which the Earth originally formed has long become separated from non-metallic components as a consequence of planetary geologic processes. Metallic nickel-iron ultimately will have been concentrated to form the earth's core. The earth can be thought of as a gigantic accumulation of meteoritic material which has been processed by melting and other geologic processes many times over. The same thing has happened with Venus and Mars, metallic iron present in the stuff of which these planets originally were made now resides in their cores. *The melting processes that accompany planetary formation and evolution effectively separate the two major components, metal and non-metal.*

Vague evidence of chondrules can be seen in this slice of a dark NWA stony meteorite. Both chondrules and clasts have been all but obliterated as a consequence of **metamorphism,** which took place sometime during the meteorites residence in solar orbit. Note the round metallic globule to the right of the penny. The white specks and globules are composed of metal (nickel-iron), a component of almost all meteorites.

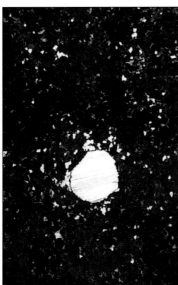

Close-up of the nickel iron globule shown in previous photo.

Highly disseminated nickel-iron (H5 chondrite) in NWA meteorite. These are some of the most distinctive and striking of NWA stony meteorites, informally referred to (here) as a "**starry night.**" (Value range E).

Slice through a large nickel-iron meteorite (Campo del Cielo). The intergrowth of nickel iron (kamacite) makes this pattern one of the very distinctive features of a sliced and etched metallic meteorite. Interlocking metal crystals like this are known as **Widmanstatten Figures**. Note the oxidization between crystals at the left. This is a bane of many meteorites because they (or at least some of their components) don't like our atmosphere, particularly its free (elemental) oxygen and its water vapor with which they chemically react. (Value range D).

Meteorites vs. Earth (Terrestrial) Rocks

Most of the rocks illustrated here formed under conditions vastly different from those which produced earth rocks. Meteorites are natural materials and as such they have an origin that reflects and is dependent upon the conditions under which they formed. Their origin is a consequence of the application of the universal laws of physics and chemistry operating under specific conditions—and in the case of most meteorites, these conditions are **those of space** and not those of a planetary surface. Their origin was not influenced by atmospheres and running water as is the case with earth rocks and (to a lesser extent) the rare SNC (Martian) and Lunar meteorites. Meteorites, being rocks, can be "read" by a geologist, and such a reading can convey information (or at least a hypothesis) regarding conditions which existed when they formed. These conditions go as far back in time as is possible to go with chunks of actual matter. In other words, meteorites (for the most part) are vastly more ancient than are most earth rocks—some having been formed **before** the planets themselves formed. Understanding the genesis of meteorites will "get you into **space**" and its astronomical past, not into the earth and its geological past as is the case with earth rocks.

Meteorite Composition

Meteorites, having come from somewhere other than the earth, might be expected to possess exotic compositions and such is the case with some of them. Certainly the metallic meteorites, the siderites, with their solid metal and high density, are in the realm of **weird rocks**. Stony meteorites also have some of the properties that might be expected to support their having come from outer space, which includes some weird chemistry. Compounds such as phosphides, hydrides, and carbides are found in some of them, compounds that are not found in planetary rocks. Various other attributes allotted to space rocks, however, like their ability to hum, fizz or glow in the dark are, I'm afraid, either the product of an overactive imagination or may be attributed to imbibing in too much alcohol or to the use of some illicit substance. It might be added however, that some meteorites do contain calcium sulfide, the mineral oldhamite, which is the material added to plastic to make "glow in the dark" objects. Most stony meteorites however, unlike metallic ones, look like plain, ordinary rocks to the uninformed. What is plain about them is that their predominant composition is of silicate minerals. Silicate minerals are also comprise most earth rocks, but **here is where the similarity ends**.

Silicate Minerals

Silicate minerals comprise a complex group of chemical compounds consisting of **some element** (or more likely elements) which are chemically combined with **aluminum**, **silicon,** and **oxygen**. The "some element" is generally iron, magnesium or calcium—the same elements common on earth. Found less commonly in meteorites are minerals of sodium or potassium, silicates of these alkaline earth elements being more characteristic of earthly minerals and the terrestrial rocks made up of them. Silicates of iron, magnesium or calcium, the silicate minerals generally found in meteorites, tend to be characterized by their formation at high temperatures and exist primarily as two major mineral groups, one known as the **pyroxenes**, the other as the **amphiboles**. Another silicate mineral, (or rather a small group of silicate minerals known as the **olivine group**) consists of iron and magnesium silicates which lack aluminum—aluminum being present however in both the amphiboles and pyroxenes. The other group of silicates found in meteorites is known as the feldspars, however only one of these, calcium feldspar, is commonly found. Most of the silicate minerals found in meteorites occur also in terrestrial rocks, particularly in igneous rocks where they sometimes form nice, collectible crystals—something usually not seen in most stony meteorites.

Augite, an iron, magnesium alumino silicate: One of the most commonly found silicates in stony meteorites is augite, but it is a mineral which usually occurs only as very small crystals. Augite is a **pyroxene**, having a chemical structure of single chains of silica tetrahedral combined with aluminum, iron, and magnesium. Pyroxenes of various types can make up much of the silicates of stony meteorites.

Hornblende is a mineral made up of iron-magnesium-calcium alumino-silicate. Hornblende is found in meteorites less frequently than is augite. When a component of a stony meteorite, it is identified only by observing the meteorite in thin section with the use of a petrographic microscope. Hornblende is an amphibole, that class of ferro-magnesian silicate minerals characterized by double chains of silica tetrahedra.

Anorthite is a calcium feldspar. It is the only feldspar commonly found in meteorites, the other feldspars being common components mainly of earth rocks. Large parts of the moon (Luna) are found to be composed of this mineral, specifically the lunar highlands.

"Medallion" pressed from liquid basalt, which came from an eruption of Mt. Vesuvius early in the twentieth century and was distributed by Wards Natural Science Establishment. Wards had these pressed with a waffle-iron-like device from basalt lava which issued from one of Vesuvius's mafic eruptions (Mount Vesuvius erupts both mafic and felsic lavas at different times). Basalt is a **mafic** igneous rock which is similar in composition (in some ways) to the silicates of many stony meteorites. The eucrites, one type of achondrite, are composed of basalt.

Olivine (known also as peridot as a gemstone) is an iron magnesium silicate usually with a beautiful olive or grass green color. The gemstone peridot consists of single olivine crystals like these. Olivine is especially characteristic of the pallasites, a category of beautiful stony-iron meteorites.

Orthoclase Feldspar: Orthoclase, a potassium aluminum silicate, is a feldspar. It is a common component of terrestrial rocks, but is not found in meteorites (or is found rarely, to say **never** with regard to natural phenomena like meteorites is dangerous as exceptions can sometimes be found, see disclaimer). As a feldspar, orthoclase, like all feldspars, is made up of a network of silica tetrahedra.

Olivine crystals in basalt: These greenish xenoliths (strange rocks) are made up of olivine. They were carried to the earth's surface by mafic lava (basalt) originating in the earth's mantle. Rock chunks at the bottom are dunite. This green rock is composed almost entirely of olivine crystals.

Granite (containing orthoclase): This pink (orthoclase rich) granite is a common terrestrial rock. It is **not** found as a meteorite. Orthoclase is a felsic mineral and felsic minerals (and the rocks which these minerals make up {like granite}) are common components of the earth's continents. Continents represent the felsic part of the earth's crust. Felsic minerals and rocks also are **not** found as components of the ocean floors, which are mafic in composition. The two most distinctive parts of the earth are fundamentally defined by the rocks which compose them—*continents=felsic rocks; ocean basins=mafic rocks.*

Orthoclase crystals (pink) are present in this peculiar granite. The black crystals in this granite are those of augite (a pyroxene). Note the rim of some mineral other than orthoclase (plagioclase?) covering the circular orthoclase crystal to the right. This granite comes from Brazil and currently is quite popular (2010) for the making of kitchen counter tops. Because of its abundant orthoclase crystals, this obviously is a typical terrestrial (earthly) igneous rock.

A terrestrial polymict breccia (it can also be considered an igneous rock) with a "meteorite-like" composition. Note, however, that two granite clasts (pink orthoclase bearing fragments) lie between a mafic clast. The majority of this rock is peridotite, an **ultra-mafic** igneous rock which originates from the **earth's mantle**. Compositionally peridotite is similar to the silicate components of stony meteorites. This peridotite was forced upward, under high pressure, from the earth's mantle along a major fault in the eastern Ozarks of Missouri (Ste. Genevieve Fault) some 300 million years ago. On its way up it picked up samples of the different rocks being traversed, which included this red, orthoclase-rich granite.

Sodium plagioclase (a feldspar) and quartz: This is a coarsely crystalline igneous rock made of earthly minerals, but having an **extraterrestrial connection**. This coarse crystalline rock is known as pegmatite; it came from the center (ground zero) of the Decaturville astroblem south of Lake of the Ozarks in Missouri. The Decaturville astroblem was formed from an asteroid hitting what is now south central Missouri some 200+ million years ago, creating a crater about ten miles in diameter. The crater has since been eroded away, but consequences of the impact are still observable, which includes the Decaturville Structure itself along with its breccias, shatter cones, and shocked quartz—all of which still exist in the form of anomalous ("strange") geology. The pegmatite from which this sample originated came from at least 1,200 feet below the surface and was carried upward as a consequence of rebound from the impact. This rebound brought up rocks from deep below the impacted surface at "ground zero," the **center of the impact**—such a rock is otherwise unusual for Missouri. At the right of the feldspar bearing pegmatite is a sample of mica schist—a metamorphic rock from the Decaturville Structure which is even rarer in Missouri than is pegmatite. Both rocks were brought up by the rebound of impact.

Meteorite Classification

There are various schemes used in classifying meteorites. Some of these are petrographic in their requirements, in that they require the making of thin sections and examination under a petrographic microscope using polarized light. Classifications used here are of a simpler, first order type—that is, they are based upon visual inspection **only**. Various first order meteorite classifications are currently used and the oldest one (but easiest to understand) is the tripartite, one consisting of **stony, iron, and stony-iron** categories. When it comes to getting to know meteorites, this is the simplest system. This coupled with references to a specific named meteorite, its name having to do with where on earth it was found, can identify and classify most meteorites, whether they are from a find or a fall. Thus Paragould Arkansas (a stony), Canyon Diablo (an iron), and Brenham Kansas (a stony iron) brings to mind a specific meteorite type, both in terms of composition and texture.

Meteorites vs. Meteor-wrongs (Pseudometeorites)

Dark, heavy rocks of various types can easily be confused with true meteorites. Various processes, both man made and natural produce "rocks" that can have the look and characteristics of space rocks. Such rocks, really can be confusing and are referred to as pseudometeorites or as **meteor-wrongs**. Pseudometeorites include a variety of natural and man made rocks, the diversity of which can be considerable. To sort out what actually is a meteorite and what is not requires someone who is an expert on meteorites and who has seen a lot of them. Someone who is able to distinguish between what can be very subtle differences in texture or composition, which at times can be difficult even for experts.

Possibly the most confusing and common man made meteor-wrong is slag—a silicate melt product generated in the process of iron and steel making. The author was once shown some rusty looking rocks found in "the middle of nowhere" in southern Missouri. Like actual meteorites, these had bits of metallic iron scattered through them and they were somewhat heavy. Most importantly, **a magnet was attracted to them**, which is a characteristic of **most real meteorites**. Upon sawing one of these open, there was metal—small globules of metallic iron, but it looked somewhat suspicious. The globules had a different signature from that found in real meteorites and it also had the subtle signature of the iron globules found in slag. In addition, there was the presence of gas bubbles (vesicles), something rarely seen in meteorites. What the finder had

come upon was the product of a nineteenth century iron smelting facility, which had been set up near some nearby iron deposits, a practice common in the early nineteenth century when Missouri was on the frontier and a supply of iron was needed. Small, wood burning smelters would be built from local rocks and these used to smelt nearby iron ore, as transportation of iron from distant sources at the time was difficult. Slag generated from these operations can still be found in places in the Ozarks, which are *still* in the "middle of nowhere." What really proved however, beyond any doubt, its being a meteor-wrong was that the specimen contained bits of carbonized wood (charcoal) derived from the wood burning smelter. Other common meteor-wrongs include various terrestrial igneous rocks, volcanic glass and various terrestrial iron minerals, all of which can mimic a true meteorite.

Ersatz fusion crust: What turned out to be a meteor-wrong was offered as a possible meteorite. The specimen has a thin, dark glassy (iron rich) skin or crust. It resembles the fusion crust of a meteorite fall (or a recent find). Such a crust is produced when the meteor is intensely heated by friction with the Earth's (or Venus's or Mars's) atmosphere.

Ersatz fusion crust—another view.

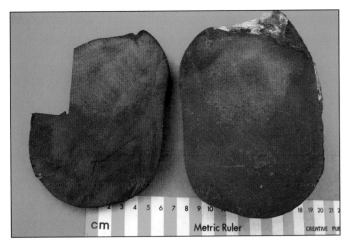

The above meteor-wrong sliced! It's made up of hard, iron-bearing sandstone or quartzite. Iron in the rock is responsible for the dark pigmented "fusion crust." This meteor-wrong turned out to be a cobble to which someone applied the intense flame of an oxyacetylene torch.

Fulgurite: When lightning strikes sand (particularly in arid climates) it can melt and fuse sand grains together, producing objects like this made of fused quartz (quartz glass). It is known as a **fulgurite** and they are found in desert regions where meteorites are also found. No meteorite looking like this has ever been found either as a fall or a find, although some terrestrial impact related glass can resemble fulgurites. (Value range F).

Obsidian, partially devitrified: Obsidian (natural glass) and unnatural glass known as slag are often confused with meteorites and considered to have an extraterrestrial origin. Glass is felsic in composition and **no meteorites** (other than tektites {which are **not really meteorites** and are **not** extraterrestrial}) are of felsic composition. Glass, for the most part, is an earthly thing! Glass also is a semi-stable material. Over spans of geologic time, glass generally devitrifies (crystallizes), becoming a finely crystalline, opaque material.

Spherical meteor-wrong: A bright meteor display (boloid) encouraging a southeastern Missouri resident to look the following day for possible meteorites that might have fallen, came up with this in a field. It's not a meteorite; rather it is a concretion covered with what resembles small craters, a seemingly perfect association for a meteorite if one doesn't take into consideration the removal of any cosmic small craters by the ablation process upon entry. *Courtesy of Bill Melton.*

Back side of spherical meteor-wrong.

Vesicular basalt: Rock containing many obvious gas bubbles like this is almost always terrestrial. Some lunar rocks (which are found as a type of rare meteorite known as a lunar) can contain vesicules, but these are rare (but highly desirable and pricey). Vesicular mafic rock, like basalt, is quite common on the earth. It is used to make roads in many places; however, pieces of basalt like this seem to suggest to many persons a rock of extraterrestrial origin. It's a very unlikely event that a vesicular chunk like this was blasted from the moon by impact. Such a specimen is most likely a meteor-wrong.

Slag with metallic iron: The iron "blebs" (white as seen here as a metallic luster will not reproduce photographically) are irregular (that is they have ameboid-like shapes) in this sliced specimen. They also somewhat resemble the shape of the mineral-filled vesicules of the previous photo. Some nickel-iron "blebs" in meteorites do have a similar shape. The black material is silicate in composition. It is a man-made igneous rock similar to basalt. This slag came from an old iron foundry and slag like this is commonly misidentified as a meteorite, that is, it is a meteor-wrong or pseudometeorite. Its metallic iron, which shows up as white in the photos, unlike that in true meteorites, contains **no** nickel.

Vesicular basalt with mineral filled gas bubbles (vesicules): Colorful silicate minerals (light green epidote), dark green (hornblende), red (orthoclase feldspar), and white (quartz) have filled vesicules in this terrestrial basalt. These filled vesicules in some way resemble chondrules, but they have an entirely different origin (and they also have a different signature, which a knowledgeable meteoritist would immediately recognize).

Two slag specimens from an iron furnace. Note the elongate trend of the metal, a phenomena rarely observed in bona fide meteorites.

Greenstone with irregular vesicules: Greenstone is metamorphosed basalt. It can comprise some of the earth's oldest rocks and is mafic in composition. The dark regions are mineral-filled vesicules but these are not spherical as are those in the previous two photos. Greenstone is mafic but meteorites with a texture (and color) like this have not been found—but some eucrites do come close.

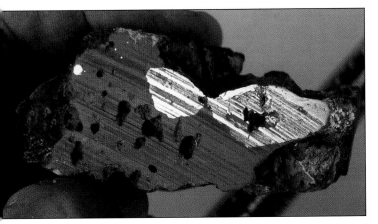

Metallic material with a flattened-out-shape in a slag meteor-wrong (or pseudometeorite): Iron in the liquid state is immiscible with the liquid silicate melt. Gravity causes both liquids to flatten out, the molten silicate melt floating as a layer above the molten iron in the ladle. Some masses of molten iron have remained in the silicate melt like those shown here while others were flattened by earth's gravity. Such flattening is normally not seen in the metallic inclusions of stony meteorites, the gravity of the parent body in which the molten material occurred being too low, only a small fraction of that of the earth. Hence, the flattened form of small iron masses like those seen here is an earthly thing and something usually not seen in stony meteorites.

Another flattened iron mass in silicate slag. Note the metallic iron filled gas bubbles in the slag. The shape of the gas bubbles imparts the shape to the metallic "blebs." Note also the unfilled bubbles to the right.

Ersatz "meteor-wrong nose cone" on the campus of Florissant Valley Community College, St. Louis. This boulder has the shape of a large meteorite sculpted by ablation upon its entry into the earth's atmosphere. No fusion crust is present and its major component is quartz, and quartz is **not** found in meteorites!

Iron Distribution in Meteorites

The earth is really a unique place, not only because of the presence of life, but also because of the composition of many of its rocks. A major category of earthly (or terrestrial) rock forming minerals are known as **felsic** minerals. Felsic rock is rock made of felsic minerals, which forms the stuff that makes up most of the earth's continents. Felsic rock is characterized by being high in silicon, sodium, and potassium. As a consequence of its composition, it has a relatively low density (that's the reason the continents rise so much higher above the more dense ocean basins). In contrast to the continents, the earth's ocean basins are made up of the other major geochemical category, a category known as **mafic** rock. Mafic rock is made up of **mafic** minerals, which are high in heavier elements like **iron**, **magnesium**, and **calcium**. Iron often imparts a greenish or black color (ferrous iron) to these minerals and also to the rocks made up of them. Disregarding the metal in meteorites (which is in a class by itself), most stony meteorites are **mafic in composition**. In fact, other than the earth's felsic continents, mafic silicate materials make up most of the rocks of the inner planets, including Earth, Mars, Venus, and Mercury, as well as the asteroids in the asteroid belt, the source of most meteorites. The moon is also primarily mafic in composition. That is the reason its

A meteor-right—the real thing! This weathered NWA stone, before being sawed and polished, looked like an ordinary rusty earth rock. To really get to know meteorites (and to really know them is to love them) one has to become familiar with the (sometimes) subtle signatures associated with them.

Another lawrencite **ugly!** This meteorite, before being cut, sat for hundreds (perhaps thousands) of years in the Sahara Desert, its lawrencite protected by being buried in a silicate mass, which is a good sealant. When the meteorite was cut, all this changed! This is one of the down sides of meteorite collecting; meteorites really don't like the water and free oxygen in our atmosphere!

It's particularly ugly, but its still nice because it's a real meteorite. This NWA meteorite, even when sliced, still looked like an ugly, ordinary (rusty), uninteresting rock. A few months after being cut, ugly, rusty looking blobs appeared. This is because,
- 1. This is a real meteorite, and
- 2. This meteorite contains lawrencite, a ferrous chloride mineral which is a mineral **not** found in terrestrial rocks. Lawrencite does not like our atmosphere and after being cut, the meteorite's interior was then exposed to the atmosphere's water and oxygen. Being a meteorite, it doesn't like either one of these (or from the "glass half full" perspective, it *does* like them and combines with them)! A chemical reaction occurred which produced both hydrochloric acid and hydrous iron oxide (the rusty blobs). Once the HCl is produced, it starts a chain reaction, producing more lawrencite and what is known as meteorite "disease" takes place. Earth rocks *rarely* behave in this "nasty" sort of way.

surface is so colorless. Mafic rocks and the landscapes made up of them tend to be grey or black like the moon's surface. Mars you might say is an exception, it's predominantly **red!** But its red color is derived from the iron of its mafic rocks, iron in this case having been oxidized (to ferric iron) by hundreds of millions of years exposure to sunlight and its ability to photo-dissociate water. It's the free oxygen derived from water that oxidizes ferrous iron to the reddish ferric iron responsible for the red color of Mars. Iron, in its oxidized (ferric) state, forms red pigments (the red of hematite), which is also the source of the red rocks and soils of the US desert southwest. Ferric oxide is the source of the Martian red color (or possibly iron peroxide in the case of Mars, which can be even a bit more intense in its reddish color).

Chemically combined iron in minerals like pyroxene and amphibole represent one type of iron occurring in meteorites. The iron in this case is combined with other elements in the form of a **chemical compound**. The other category of iron-occurrence, one totally **separate and different** from the ferrous iron compounds, is **metallic iron**. These two occurrences of iron represent **two different and distinct categories** of an element that is so important in meteorites. **This duo-occurrence of iron is found only** (but see the disclaimer in the glossary) **in meteorites**.

Stony meteorites, being representative of silicate materials of the Solar System, are mafic in composition and therefore are prevailingly dark (or grey), with few warm colors prevailing, unlike the situation on earth where warm colors (associated with rocks like granite, rhyolite, and the sedimentary rocks derived from them) prevail. **As a general rule, felsic rocks and minerals are warm colored, mafic ones are black, grey or somber colored**, viz. greenish or bluish.

Note the even distribution of metal in this meteorite. A zero or very low gravity field is responsible for such an even distribution.

Chondrites can barely be seen in this meteorite slab. They were once present, but metamorphism caused from both shock and heating all but obliterated them.

Mars rock: The red color of Mars is due to ferric iron oxide (possibly, in part, ferric peroxides). This red color comes from ferrous iron on the Martian surface being oxidized over long spans of geologic time by oxygen produced (probably) from dissociation of water. At the right is a very small Martian meteorite fragment. Its iron is in the ferrous oxidization state characteristic of most mafic materials. These Martian Meteorite "sets" were sold at Tucson in 2006. (Value range E).

The Age of Meteorites

Meteorites are the most ancient things geological! They are also the most ancient things one can touch and directly investigate! Some meteorites (or parts of them) even predate the formation of the Solar System! Planetary accretion took place some five billion years ago, and the most ancient "planetary" rocks, which have been dated, are moon rocks that came from the lunar highlands. These lunar rocks are dated at 4.5 billion years. Nickel-iron meteorites, formed in the interior of asteroids, also date back to 4.5 billion years, so these are also older than any (known) earth rocks. The oldest known earth rocks date at (almost) 4.0 billion years; they form the Isua Series of Greenland, and rocks slightly younger than this are associated with greenstone belts of the Canadian Shield and western Australia.

The oldest rocks of all are the **chondritic meteorites**, whole rock radiometric age dates (geochrons) on them can give dates of around 5.0 billion years. This age, as well as the composition of these meteorites, represent a form of material which is in a **primitive condition**. Primitive compared to the stuff of metallic meteorites, which was derived from the melting and partial processing of the **more primitive chondritic material**. Chondritic matter is believed also to have been the parent material responsible for forming the terrestrial planets—Mercury, Venus, Earth & Luna, and Mars. But, as with the Earth, material on these planets has been so extensively processed by planetary activity that it has little resemblance to the primitive material found in chondritic meteorites. Individual components of these meteorites (like the chondrules themselves) can give even greater ages—six billion years—a date which predates the formation of the Solar System by over a billion years. This age is also consistent with the most accepted origin of the chondrules—small spheres formed in a zero gravity field from matter thrown off by the "dying" stages of various types of stars. These small spheres clumped or accreted together with other matter to produce material, which later clustered forming the solar nebula, which in turn became the material that formed the planets and the sun. Thus the chondrules of chondritic meteorites represent the most **primitive** as well as the **oldest** original form of matter that you can actually observe, touch, and analyze.

The ages of meteorites are determined using radiometric age dating techniques (geochronology). Whole rock age dates for chondritic meteorites cluster around 5.0 billion years. Age dates of individual chondrules can be greater. Age dates greater than 6.0 billion years would suggest that such a meteorite came from more ancient material from another part of the Milky Way Galaxy. Such a greater age date has not been found, but is expected by some meteorite workers as only a small sampling of chondritic meteorites have been age dated out of the large number known to exist.

Meteorites from Outside the Solar System

As far as is presently known, all meteorites (which have been documented) come from within the Solar System, most originating from the asteroid belt. As mentioned elsewhere, rare meteorites are known which came from the surface of Mars and the Moon, which are known as Martian and Lunar meteorites. With regard to the chondritic meteorites, and especially the chondrules in them, issues regarding their ultimate place of origin becomes muddy and difficult to interpret. (Meteorite workers refer to them as being "tight lipped.") Chondritic meteorites, for the most part, come from the Asteroid Belt, so in a sense they are part of the Solar System—but, the **components of these meteorites**, especially the **chondrules** as well as a few other components in them (like very minute diamonds), formed **prior** to the **origin of the planets** and the sun (and hence formed prior to the origin of the Solar System) and have not been altered since they formed, a time some 6 billion years ago in the case of some chondrules and the small diamonds. In a way then, chondrules (and by inference chondritic meteorites) represent something which formed outside the Solar System as the chondrules themselves were formed before the planets and the sun formed. There is also some evidence that chondrules and the fragmental matter between them may represent matter gathered from a sizeable part of the Milky Way Galaxy. Others hypothesize that this ancient matter formed from a limited and discreet group of stars that went nova in proximity to what would eventually evolve into the Solar Nebula.

Extra Solar Meteorites

On the matter of possible extra-solar meteorites, a few puzzling, yet reliable occurrences have been considered. Probably the best known is the Eaton Colorado fall associated with the meteorite collecting of Harvey H. Nininger. Eaton is a metallic meteorite of a puzzling composition. Meteorites of extra solar origin might be detected by having puzzling isotope ratios or by giving age dates considerably greater than the 5.5 billion year age date of the chondrules in chondrites.

Isotope ratios in chondrules have been found to vary somewhat from those atomic weights found for the same elements in terrestrial rocks—terrestrial or earthly material representing an **average** or a **mean distribution** of the various isotopes of a particular element. Extra-solar matter might be expected to vary in its isotopic ratios if the matter which formed the Solar System was derived from multiple sources (five novas and two super novas being one suggestion). A few meteorites (some pallasites) also give spurious isotope ratios for the nickel and silica in them. The probability of any matter coming from another part of the Milky Way Galaxy and making it into the Solar System, considering the vastness of space, is quite low. (It is even lower for such a rare item to intersect the earth's orbit and to survive and be found as a meteorite). These low probabilities, however, are increased considerably with the passage of vast periods of time (mega-time), the time scale on which sun-orbiting material, the earth and the rest of the planets "operate." An extra-solar meteorite falling to the earth is an improbable event, but *given enough time an improbable event can become probable*.

A very low probability exists for a specific meteorite coming from outside the Solar System; however vast numbers of meteorites have been coming onto the earth for millions of years. The number of NWA meteorites alone found during the past few years is itself staggering; little is known or has been done with isotope ratios or age dates of most of these. Spurious meteorites are also known as finds from the geologic past—the Lake Murray meteorite of Oklahoma or the meteoritic iron found in Ordovician limestones of Sweden (some 440 million years old) being examples and these and other odd occurrences might well be examined more closely with respect to isotopic ratios.

"Starry Night:" These H6 meteorites show distribution of small metallic specks which somewhat resemble the stars of a clear, star studded sky—an appropriate pattern for a meteorite. Again the assemblage of the components of this meteorite was in a zero gravity field, which was responsible for this diffusion of metallic material. Absence of a strong gravity field is the same reason that stars are diffused the way they are. If the force of gravity was greater in some parts of the Milky Way, the stars would then clump together.

Terrestrial Impact Sites—
Winslow Crater,
Arizona, and Astroblems

Extraterrestrial objects have been impacting the earth since it was formed, impacts apparently becoming progressively less frequent as the earth and Solar System ages over geological time. The Winslow Crater of northern Arizona is one of the most recent of such craters on the earth's surface. Older impact sites usually have the actual crater removed by weathering and erosion; the existence of what was once a large crater now being represented by the impact's effect on local geology. An impact site where the crater itself has been removed by weathering and erosion (but impact effects on the rocks are still evident) is known as an "astroblem." There are about 170 astroblems of extraterrestrial origin currently known on the Earth.

A specimen of Canyon Diablo etched to show modified Widmanstatten Figures. The explosion resulting from impact of the Winslow meteor produced such a large shock wave that the crystalline pattern in surviving parts of the meteor were obliterated by impact metamorphism. As can be seen, only a small part of the pattern of interlocking crystals, which originally formed when the parent mass of metal cooled, has been preserved.

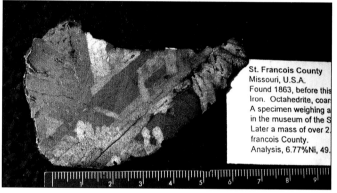

St. Francois County Missouri, U.S.A. Found 1863, before this Iron. Octahedrite, coar A specimen weighing a in the museum of the S Later a mass of over 2 francois County. Analysis, 6.77%Ni, 49.

St. Francois Missouri siderite: Here is a meteorite identical to Canyon Diablo, but one in which the Widmanstatten Figures **have not** been affected from shock metamorphism. The parent meteor was very much smaller than the one that formed the Winslow Crater so that no explosion or large crater was formed upon its impact. With such a small impact, the Widmanstatten Figures remained intact.

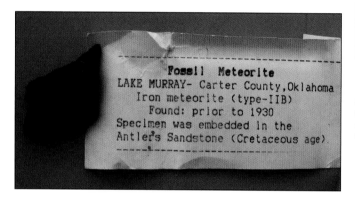

Fossil Meteorite
LAKE MURRAY- Carter County, Oklahoma
Iron meteorite (type-IIB)
Found: prior to 1930
Specimen was embedded in the
Antler's Sandstone (Cretaceous age).

Oxide crust from a "fossil" meteorite: A few nickel-iron meteorites have been found that are so called "fossil meteorites"—meteorites that fell to the earth in the geologic past and have been preserved in sedimentary rock. One of these was found embedded in Cretaceous age sandstone (approximately 75 million years old) in Oklahoma in the 1930s. Here is a fragment of its thick oxide crust formed from millions of years of terrestrial weathering.

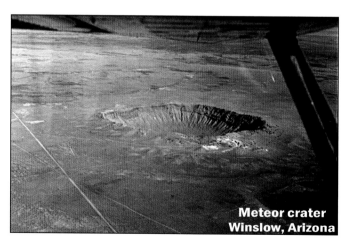

Meteor crater
Winslow, Arizona

Winslow Arizona or Barringer Meteor Crater: A small asteroid (or large meteor) entered the earth's atmosphere some 45,000 years ago and impacted northern Arizona to make this crater. High velocity impact specifically produced this crater, a crater generated from the huge amount of energy released upon impact. The meteorite itself was mostly vaporized by the resulting explosion, but portions of it were spread around the crater for at least a six mile radius. Fragments from this impact are known as Canyon Diablo meteorites after a small draw by that name where they were first found.

Meteorites, Yet Another Look!

The objects depicted here are real, and they came from space! Current culture in the US seems unable to distinguish between the world of reality and fantasy. Yes, moon rocks and meteorites don't hum, buzz or even give off noxious gasses. They are not radioactive (anymore) and they don't even mysteriously glow. On initial examination they look like ordinary rocks. But like the images of the surface of Mars or those of deep space as seen through the Hubble telescope, meteorites are real stuff from space and not make-believe fabrications or Hollywood manipulations. As with other geologic objects, meteorites give information on some pretty interesting stuff, information revealing the environment and conditions of deep space (nebula) where they formed, for example. A good geologist can read them, extracting from them what otherwise would be privy information. Keep in mind that meteorites are the only space collectible that is really *"out of this world."*

To observe the revealing interior of a meteorite requires that it be cut (or sectioned), and the most obvious, first order feature to be observed, either with the unaided eye or with a hand lens, is the condition of iron present in the meteorite. Iron present in meteorites comes in **two** fundamental forms: **metallic iron** (usually alloyed with nickel) and **iron compounds**, usually silicates. Iron compounds occur in two forms: **divalent** (**ferrous**) or **trivalent** (**ferric**) iron. Regarding the metallic iron in meteorites, this form of iron doesn't like our atmosphere; it rusts. From a meteorite's point of view, our atmosphere is a strange and noxious one containing a component which, if you were a meteorite, would find especially obnoxious, **free or elemental oxygen**! The twenty-one percent oxygen of the earth's atmosphere is really an anomaly; no other planet in the Solar System has this and probably the planets being detected around stars in other parts of the Milky Way Galaxy also lack it. Elemental (free) oxygen is a chemically reactive material; it normally doesn't stay around for a very long time, even if it were present. Rather, it combines with whatever other elements are available as it has done to produce the ferric oxide (red color) of the surface of Mars. The presence of such a reactive element in the earth's atmosphere is of course a consequence of that "**weird**" phenomena found on planet earth, which is **life**. Life, particularly in the form of photosynthetic life as with cyanobacteria, algae, mosses, phytoplankton, and terrestrial vegetation (land plants), has been pumping this highly reactive element (and noxious gas from a meteorite's point of view) into the earth's atmosphere for at least two billion years. This condition has not been a favorable one for meteorites—not only with the elemental iron in them but also with respect to their divalent (ferrous) iron.

As mentioned above, iron comes in two forms; these are:

1. **Metallic Iron**. This is a form of iron in which the iron atoms are quite happy; they all have the same number of electrons as they do protons and the outer electrons are able to wander through the iron quite freely. Metallic iron is a common material in our Solar System and probably in other star systems as well. Stars and the stuff blown out from stars in novas and supernovas make iron (element # 26). In fact the nuclear reactions *going on in stars are "programmed" to make element number 26 in vast quantities.* Metallic iron, of course, is a metal! It is shiny, has a metallic luster, and conducts an electrical current quite well because of those wandering electrons.

2a. The other form of iron exists as **compounds**—two groups of which exist—the one found most commonly in meteorites is **ferrous iron**. With this form of iron, two electrons become removed from each iron atom so that for the atom to exist and still be "happy" ("we always strive for happy atoms"), iron atoms have to join-up or combine with some other element and that element often is oxygen; *ferrous iron* compounds in meteorites and in earth rocks are generally greenish, black or grey in color.

2b. The other form of iron found as compounds is known as ferric iron. Ferric iron represents iron atoms that have lost three electrons from each atom so that the atoms need to combine with something and again this something often is oxygen. Compounds of this form of iron are usually yellowish, brown or red, depending upon whether there is water associated with the iron oxide. Ferric iron compounds are the ones common on earth and they produce the reds and yellows of "earth colors" found in many soils and rocks. Ferric iron is also responsible for the red color of Mars, where the free or elemental oxygen has come (probably) not from photosynthesis (as is the case on the Earth), but from a process known as photo-dissociation, this being the very slow break down of water into its component hydrogen and oxygen from ultraviolet waves originating from the Sun.

Few extraterrestrial materials contain ferric iron, the one exception being the reddish regolith (loose material on a planetary surface) found on Mars. Meteorites containing ferric iron generally are finds that have been acquired during the process of weathering (terrestrialization) while the meteorite lay on the earth's surface. Meteorites that have reacted with the earth's oxygen and water are those that have had a long residency on the earth. Examples of such meteorites are the rusty looking finds.

Meteorites as Collectibles

Meteorites represent a unique category of collectibles; they represent the only collectible which is actually from **another world**. They also represent a geologic collectible that appears to hold its value or even appreciate in value with the passage of time. The author, whose formal education is in geology, was led to believe as a student, that these extraterrestrial objects were quite rare and exotic and to a degree this is true. However they are not so rare as to be inaccessible. The availability of meteorites has, particularly during the past twenty years, become greater for a variety of reasons. One of these being a greater interest in them, and interest has increased in what forty years ago were considered to be exotic and somewhat inconsequential objects. This increased interest has resulted in more persons looking for them and more of them have thus been found, sometimes significantly more. The "flood" of meteorites from dry deserts of Africa has particularly increased their availability, to the point where a person currently (2010) can acquire a reasonably good collection of meteorites with an outlay of only a moderate amount of monetary resources.

The values given for many of the specimens, especially northwest African meteorites (NWA's) are highly negotiable; the value of NWA's, as a consequence of the "flood" of them that has taken place during the past few years, has resulted in a great deal of instability to the value of these meteorites.

Value range used in this book
    A. $1,000-$1,500+
    B. $500-$1,000
    C. $250-$500
    D. $100-$250
    E. $50-$100
    F. $25-$50
    G. $10-$25
    H. $1.00-$10.00

Glossary, Chapter One

Ablation: The heating of a material entering the atmosphere with its surface becoming very hot from friction so that surface material vaporizes, forming a plasma.

Alkaline Elements: Elements IA on the periodic table, that is lithium, sodium, potassium, and other elements having one valent electron. Of these, only sodium is found in abundance in meteorites; the others, including potassium, being uncommon.

Atomic Number: The number of protons in an atoms nucleus. The arrangement of elements according to their atomic number is known as the periodic table.

Atomic Weight: The number of protons and neutrons in the nucleus of an atom. Some atoms with the same number of protons (atomic number) have different numbers of neutrons. This variance in the number of neutrons in an atom reflects conditions in the star that originally produced the particular atom.

Breccia and Brecciation: Rock broken up by some process—with meteorites and other extraterrestrial phenomena, that process usually is impact. Breccias are made up of angular fragments (clasts) and matrix that may be rock flour or dust coming from various sources, including the impacting meteorite.

Clasts: Angular fragments making up a breccia. These fragments may be from the same source (monomict breccias) or from different sources, including fragments of the impacting meteorite itself (polymict breccia).

Coesite: A form of shocked quartz produced from high velocity impact. The presence of coesite associated with an impact structure is considered as a strong argument for a structure's extraterrestrial origin. Coesite is a polymorph of quartz, another rarer polymorph is known as stishovite. Both of these minerals have to be identified by X-ray diffraction.

Compound (chemical): A substance in which the elements composing it are in chemical combination involving the presence of chemical bonds. Chemical compounds have properties that are usually entirely different from those of the elements of which they are made.

Cone-in-cone Structures: Cone-in-cone structures (often with the cones nested) develop in terrestrial sedimentary rocks. Cone-in-cone structures can easily be confused with shatter cones. A person familiar with both of these structures can easily tell the difference between them.

Cosmic Velocity: The speed (or velocity) of an extraterrestrial object (relative to the earth's velocity) before it is slowed down by atmospheric friction.

Disclaimer: With natural phenomena there is usually an exception to most general statements. The statement made that metallic iron is not found in earthly rocks is true; however, there are a few rare exceptions to this, like the mineral Josephinite of Oregon and the elemental iron of Greenland, which formed from being naturally smelted from an iron rich basaltic magma in its encounter with a coal seam.

Geochron: A numerical age date (involving geologic time) determined on the basis of the ratio of the amounts of a parent radioactive element present compared to the amount present of its decay product—the more decay product, the greater the age of the sample.

Isotope: An element of the same atomic number but differing atomic weight (or mass).

Isotope Ratios: A ratio between one form of an element (isotope) and another of the same element but with more neutrons. Different stellar events (like a nova and a supernova) can produce not only different amounts of the same element, but also different amounts of isotopes of the same element. Earthly and other planetary materials give isotope ratios which (in most cases) are consistently the same and represent averages of the various isotopes present. Thus, the atomic weight of an earthly element (iron) might be 55.85, which is an average or mean of the different isotopes of that element present. In primitive meteorites, viz. chondrites, isotope ratios may vary. The first meteorite might contain iron with an average atomic weight of 55.92, and the second one with 55.85, the first specimen reflecting the presence of a greater amount of the heavier isotope Fe-56 in it and less Fe-56 in the second one. Different isotope ratios are believed to reflect an origin of the iron composing the two different meteorites, having formed in two different stars.

Lunar Meteorites: Rock can be blasted from the moon's surface by high velocity impact (spallation). An escape velocity of only .5 Km/sec is required to leave the moon's gravity and this is easily achieved by the high velocities produced by impact.

Metal and Non-metal: The metallic state of an element has the same number of electrons as protons and the electrons are allowed to move readily from one atom to another, thus metals conduct an electric current. Metals also have a metallic luster (which cannot be photographically duplicated—it comes out white or grey) and they are ductile or malleable. Non-metals lack these properties.

Meteoroid: A rock-sized object in space. When a meteoroid gets to be a kilometer or more in diameter it is then known as an asteroid (although a small one). No distinct size range separates a meteoroid from an asteroid. The two terms evolved separately and its only that an asteroid is much larger than a meteoroid.

Meteor: A space rock (meteoroid) which enters the earth's upper atmosphere and becomes intensely heated by friction. Light and other visual effects accompany a meteor, which may (or may not) be entirely destroyed by vaporization or ablation upon entry into the atmosphere.

Meteorite: A space rock (meteoroid) that survived the intense conditions of entry into the earth's atmosphere and was not destroyed by the intense friction of entry so that it impacts and is recovered. When a meteor impacts with the earth's surface, it becomes a meteorite.

Meteorite Market: A small market produced for meteorites by both the interests of science and collectors. Often a meteorite when found or seen to fall will be cut in half, one half being retained by the collector or institution that acquired it. The other half will be traded or sold, often to be cut into small slices that then may be traded or sold to other collectors. The trade and purchase of these small slices, as well as complete meteorites (often from well collected and prolific fields), constitutes the market.

Mineral: A naturally occurring element or chemical compound of inorganic origin. Some 3,800 minerals are known (depending upon who does the counting); some of these minerals are known to occur exclusively in meteorites, like hydrides, carbides, and some sulfides.

Periodic Table of the Elements: An arrangement of the elements in the order of the number of protons in the nucleus (atomic number). The periodic part of the table resides in the repetition of certain chemical properties as higher atomic numbers are reached.

Plasma: A fourth state of matter. The process of meteorite (and spacecraft) ablation produces this highly energetic state of matter, which is somewhat like, but much more energetic than, a gas. This is a forth state of matter—beyond a gas is a plasma.

Regmaglypt: A deep pit or cavity on the exterior of meteorites produced by ablation or certain minerals in the meteorite as it passes through the earth's atmosphere.

Residency Time: The time that a meteorite has been on the earth or another planetary surface. With earthly finds, long residency times generally result in a meteorite appearing brown and rusty. Such a rusty appearance is from the formation of ferric iron compounds like limonite and goethite, compounds absent in fresh meteorites.

Shatter Cones: Cone-shaped structures 2-40cm in length having a distinctive signature and produced in rock from high velocity impact. The presence of shatter cones is one of the critical criteria supporting an extraterrestrial origin of astroblems and craters.

Solar Nebula: A once existing nebula which condensed some 5+ billion years ago and gave rise to the Solar System with its central star and cadre of planets.

Shocked Quartz: A form of the mineral quartz in which the composing silica tetrahedra are "skewed" as a consequence of high velocity impact when the shock waves "add up" to produce a super wave known as a solutron. Shocked quartz, especially coesite, found in earthly astroblems, can be associated with shatter cones.

Silicate Minerals: Minerals composed of the element silicon and oxygen, plus a number of other elements. Generally these "other" elements are aluminum (which sort of substitutes for silicon in the crystalline lattice), sodium, calcium, iron, and magnesium. Potassium is also commonly present in earth rocks, but potassium silicates are rare in meteorites.

Silicon: Element number fourteen on the periodic table. Silicon is one of the most common elements in the Solar System where it and iron are key elements composing much of the mass of the inner planets.

Thin Section: A thinly cut small rock (or meteorite) slice glued to a glass slide and then ground down until it is so thin that it transmits light. Such a "slide" can then be examined by transmitted light through a microscope. (Often a petrographic (rock-looking) microscope.)

Texture: The pattern of crystals of different sizes or types in a meteorite or igneous rock.

Vesicules: Gas bubbles in igneous rock. Vesicules are rare in meteorites, although Lunar and Martian basalts, as well as some eucrites, exhibit them.

Whole Rock Radiometric Age: Radiometric age dating is based upon ratios between radioactive elements and their decay products—the greater the amount of decay product of a specific radionuclide, the greater the age of the sample, and conversely the smaller the amount of decay product, the younger the sample. As meteorites (as in a breccia or a chondrite) are often made up of various components, each having a different origin and hence a different age, a radioactive age date for an entire meteorite will give an age representative of the **average** or **mean ages** of the **different components in the meteorite**. A radioactive age date for an individual clast (or chondrule) might be considerably greater than a "whole rock" age because these formed at an earlier time. Thus a whole rock radiometric age or geochron *decay product to parent radionuclide ratio* might be ½ (.5). Individual components of a meteorite like a clast or a chondrule on the other hand, might have a greater decay product to parent nuclide ratio like ¾ (.75), hence would be older than the age obtained from using the "whole rock" meteorite as the chondrule or clast formed **before** the rest of the material did and **therefore had a greater amount of time to accumulate more decay product**.

Xenoliths: "Strange rocks" incorporated into and carried by magma. Xenoliths are somewhat like clasts; however, a clast is a rock fragment in a breccia. A xenolith is a rock fragment (sometimes quite large) plucked from below a planet's surface by magma and carried by this magma at or near the surface. Xenoliths are always found in igneous rocks, which includes achondritic meteorites that are extraterrestrial igneous rocks.

Chapter Two
Metallic Meteorites

Origin of Metallic Meteorites

The most primitive (and oldest) meteorites are known as chondritic meteorites. They consist of a mix of fine granular and/or crystalline material through which are dispersed small spheres 2-8 mm in diameter, some of which, upon closer examination, appear to have once been molten and then slowly crystallized. Both the granular material and the small spheres (known as chondrules) are composed predominantly of silicates, many of which are an iron, magnesium, aluminum, silicate of variable composition known as **pyroxene.** The other component of these primitive forms of matter is **metal,** which is present in an **elemental state**, mainly metallic iron and nickel. In chondritic meteorites the metallic component is separate from the silicate, the former usually being present as sub-spherical "blobs" (or blebs) scattered **throughout the meteorite.** The other occurrence of metal in chondritic meteorites is in the form of small specks disseminated or scattered throughout the meteorite. The small spheres, the chondrules, were formed in a zero gravity field in a part of space where matter could be suspended and was not attracted by the gravity of any large body. Chondrules were formed from matter believed to have been thrown off from a star in the last stages of its "life," when it was either a nova or a supernova. Chondrules are the most easily observable parts of the primitive parts of stony meteorites. The last stage in a star's "life" fuses lighter elements like helium into heavier ones, which in a **nova** can produce elements up to and including iron. A nova, an exothermic nuclear reaction involving fusion, fuses lighter elements into heavier ones, throwing off, in the process, a stream of matter. This matter may then condense to form or add to a nebula. It was in such a nebula that matter accumulated that comprises much of the mass of the Solar System (and meteorites). *Chondritic meteorites, especially their chondrules, were privy to the early-most stages of formation of the Solar System.*

Those elements not created in a nova are the heavy ones, which include long-lived radioactive elements like uranium and some of the trans-uranium elements. These are produced in an exploding star of **vastly greater energy output known as a supernova.**

Carbonaceous chondrite, Allende, Mexico. Type I (CV3) carbonaceous chondrite. The numerous circular objects seen in this meteorite slice are chondrules. This is a fresh chondritic meteorite that fell in 1969 near Pueblo de Allende, Chihuahua, Mexico. Chondrules are some of the oldest and most primitive components of meteorites. These meteorites contain little metal and are the farthest types from metallic meteorites as possible. Whereas metallic meteorites were formed by cooling of metallic nickel-iron from a hot, molten state, carbonaceous chondrites were never heated above 30 degrees Celsius; if heated higher than this, their organic components would have been lost. (Value range C).

Another slab of Allende showing the abundance of chondrules. Carbonaceous chondrites are believed to (possibly) come from the cores of comets.

NWA meteorite with numerous chondrules.

Saratov, fall, 1918. This very friable (crumbly) meteorite has its chondrules falling out quite easily. Note some of them as the small spheres to the left. Heating with its consequent changes would have made this meteorite less friable and its chondrules less prone to separate.

heat, less able to dissipate in a larger mass than in a smaller one, turned the mass hot enough to become incandescent and melt. Depending upon the mass of the clump and the intensity of its radioactivity, some of these clumps became the size of a state. When a mass became "state sized," gravitational forces overcame the forces of chemical bonds and the mass assumed a spherical shape—it would then become a planetoid. Concentrated mass concentrated the heat of radioactive decay, allowing the mass to reach incandescence, thus melting it. As metal and silicate materials are immiscible in a liquid state, the two components then would become separated from each other when melted, the more dense metallic nickel-iron metal going to the center of the planetoid and the lighter silicate material forming a shell around the metal. Such planetoids might have joined together as they did in specific orbits around the protosun to become a protoplanet. Planetoids could also remain in orbit around the sun, as many did, especially those in what today is known as the Asteroid Belt and still exist—all-be-it sometimes in a fragmented form.

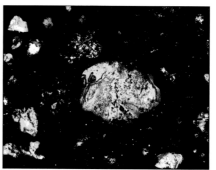

A small, irregular, clumped mass of nebular material (planetismal), which has reached bright yellow incandescence (~2,800 degrees F) from the heat of radioactivity coming from radioactive decay of short-lived radioisotopes. *Artwork by William F. Brownfield.*

Matter thrown off by a nova and a supernova was originally rich in radioactive elements (radioisotopes), most of which are short-lived and therefore were **highly radioactive**. Radioactive decay produces considerable amounts of heat and this heat often was great enough to melt the chondrules as they formed or to melt them shortly after they formed. As this short-lived radioactive matter decayed and heat diminished, the chondrules cooled and crystallized. Gradually the dispersed matter of the nebula, chondrites and all, aggregated or clumped together to produce larger accumulations—and the larger the clump, the greater the gravitational attraction. Large clumps, known as planetismals, not only had greater gravitational attraction, but **still had some short-lived radioactive matter in them**, and with the matter being more massed together, the heat of radioactivity was concentrated. This

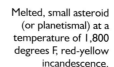

Melted, small asteroid (or planetismal) at a temperature of 1,800 degrees F, red-yellow incandescence.

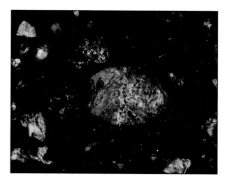

Smaller masses when accreted, if they did not achieve a mass great enough to become a planetoid, would retain an irregular, rock-like shape; these are known as planetesimals. Heating would still occur in these but because of their

smaller mass, they **did not become hot enough to totally melt**. Chondrules in these irregularly shaped masses might be changed by the heat (thermally metamorphosed) but they did not melt—thus they were preserved in this primitive condition as some of the most ancient objects which still exist today. The larger protoplanets, on the other hand, completely melted from the heat of radioactivity with the metal in them separating from the lighter silicates. The dense, liquid metal ended up in the protoplanet's interior, producing a metallic core, and the lighter silicates forming layers of lower density around this core.

Metal globules in a slab of a stony meteorite in which the chondrules have been almost obliterated (metamorphosed) by heat. The white globules and specks are nickel iron. This is a typical stony meteorite with a normal amount of iron, an H-4 chondrite. Note the elongate mass of metal to the left of the middle and also the armored chondrule below it, the metal coat of the chondrule preserving its outline. (Value range E).

The small asteroid cooled to red incandescence (1,300 degrees F.)

Cooled to black, it now radiates in the infrared, but produces no more visible light. The planetismal can be seen by reflected light of the protosun and by light given off by larger incandescent bodies, some still white hot and entirely molten.

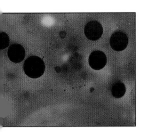

Small, cooled planetoids, which are now black (no more incandescence), formed early in the solar nebula when short-lived radioactive material was abundant enough to produce total melting and formation of spheres. At the bottom is part of a protoplanet of much greater mass, which is still incandescent and where metal of parent chondritic meteorites has contributed molten metal to form a molten, liquid core. With this larger body, because of its greater mass, a greater amount of heat was retained and the liquid metal and silicate material would completely separate from each other. **In this manner the metallic meteorites formed—the metal becoming the core of a protoplanet**. Molten silicates became mantle and crustal material cooled to form meteorites known as diogenites and eucrites respectively.

Portion of a slab of an altered chondrite (H2). Note the globule of nickel-iron at the right. Chondrules in this meteorite have been almost entirely wiped out by thermal or impact metamorphism. If this mass had melted, its metal would have mixed with other liquid metal and joined the core of a planetoid. When this planetoid (now an asteroid) was later impacted and broken, the metal would then be part of a metallic meteorite (also known as a siderite). The vertical arc in the center is a consequence of sawing the meteorite slab.

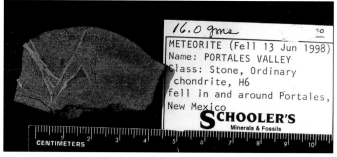

Portales Valley. An H6 chondrite with iron filled fractures. The original mass accreted some 5.5 billion years ago. The cracks appear to have been produced on impact with another object when this was part of a planetoid—the interior of which was still molten when impact occurred, allowing molten metal to infiltrate into resulting cracks formed from impact. (Value range F).

An H5 (starry night) meteorite which has highly disseminated metal scattered through a silicate matrix. This is the original distribution of the metal. If this mass had melted, it would have contributed about half of its weight (in metal) to the metallic core of a planetoid. (Value range F).

Protoplanets or planetoids having cooled after most of their radioactive material decayed, periodically impacted each other. Now being cooled and thus hard, resistant, and brittle, they shattered or fragmented to form parts of the Asteroid Belt, the source of most meteorites, including the metallic meteorites that are fragments of the cores of once existing protoplanets.

Metal present as globules (see left of center) and as irregular blebs. NWA specimen.

Toluca, Mexico: A large number of siderites have been collected south of Mexico City for over two hundred years. These meteorites were first used as material for making weapons and implements by the Aztecs. Metallic meteorites were known to ancient peoples and utilized in various ways as both implements and as sacred objects.

St. Francois, Missouri: A typical metallic meteorite with well developed and well preserved Widmanstatten figures, a feature characteristic of nickel-iron meteorites. A mass like this slowly crystallized in the interior (core) of a planetoid. The planetoid later was impacted by another one to become a smaller object (probably in the Asteroid Belt). Much later, a fragment of this was knocked off, eventually came toward the earth, and fell in what would become Missouri's, St. Francois County. This meteorite was found near Flat River, Missouri, a town now known as Park Hills in what is now known as Missouri's old lead belt.

St. Francois Missouri: Same specimen as above, but positioned so that the Widmanstatten figures are not prominent.

Gibeon: A polished and etched surface of a fine octahedrite showing well developed Widmanstatten figures. This is a widely distributed and well known nickel-iron meteorite. (Value range D).

Gibeon: Another etched slab of this widely distributed fine octahedrite.

Muonionalusta meteorite, Norrbotten, Sweden: (Left) A find that was located using a metal detector; (Right) etched slab showing interesting Widmanstatten figures. (Value range F).

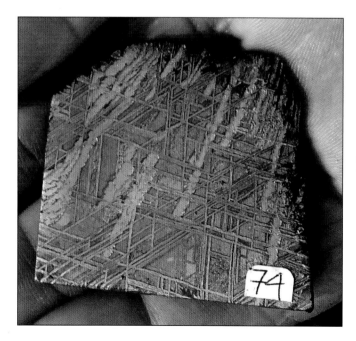

Canyon Diablo, Arizona: Canyon Diablo meteorites are fragments of the impacting asteroid that made Meteor Crater in northern Arizona. The Widmanstatten figures in it are usually vague. This is because they were destroyed by impact metamorphism from the high energy impact that made Meteor Crater. (Value range E).

Canyon Diablo: A small specimen with more of the octahedral crystals showing. These were not subjected to as intense a shock as were those specimens found nearer the crater's rim. The specimen has been encased in plastic to retard oxidization; however, oxidation and hydration still has taken place. (Value range F).

Closer view of Widmanstatten figures of Norrbotten, Sweden, meteorite. Note the triangular pattern, a characteristic of sliced and etched octahedrites.

A group of relatively large meteorites from the Sikote-Alin fall of 1947. They resemble pieces of shrapnel. (Value range D for group).

Another group of shrapnel-like Sikote-Alin meteorites.

Sikote-Alin, Russia (USSR): Vast numbers of these shrapnel-like metallic meteorites pelted the ground, producing small impact craters on Feb. 12, 1947, in the former Soviet Union. The fall wiped out the branches and needles on trees in a heavily forested area. Its visual effects were well documented.

Artwork done just after the 1947 Sikote-Alin fall as shown on a 1957 Russian postage stamp, which shows the vapor cloud from the 1947 fall of a metallic meteorite swarm.

40

Natan, China: Octahedrite. Note how this somewhat weathered specimen breaks along crystal boundaries. Natan is one of the most widely distributed of coarse octahedrites—numerous specimens having come through the Tucson, Arizona, show. (Value range F).

Octahedral cleavage fragments showing octahedral crystal pattern (equilateral triangles) on some pieces. Natan is a weathered siderite and crystalline fragments like this have been distributed widely among museums and collectors.

A large octahedral crystal of Natan: This mass separated from a larger mass along the boundaries of the crystals found in most nickel-iron meteorites. Octahedral-shaped crystals like this are why these, the most frequently found siderites, are known as octahedrites. Note the face on the octahedral cleavage mass, which also shows octahedral structure (equilateral triangles). (Value range D).

Another group of Natan meteorites: A large number of these meteorites appeared at the 2003 Tucson show. These meteorites were originally gathered from a weathered, strewn field during the cultural revolution in China to be used as iron ore designated for Chinese steel furnaces.

Slice of Campo del Cielo with "meteorite disease:" A serpentine "trail" of hydrous iron oxide covers parts of this slab—an example of the fact that many meteorites, including metallic ones, don't like our atmosphere (or perhaps like it too much, so that they want to chemically combine with it)! Campo's are one of the most readily available siderites—they come from a strewn field in Argentina, which is **quite large and extensive**. (Value range F).

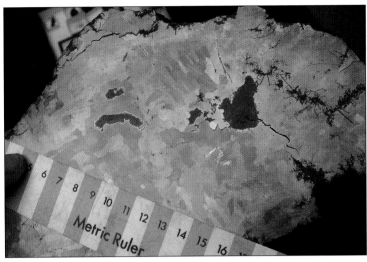

Campo del Cielo: Back side of the previously shown specimen. Note the change in shape of the graphite nodule.

Large slice of Campo del Cielo showing Widmanstatten Figures and graphite nodules (black areas). The graphite nodules are surrounded by troilite (ferrous sulfide), which shows up as slightly less than dark black. Surrounding this is a thin rim of schriebersite, an iron-nickel phosphide. (Value range D).

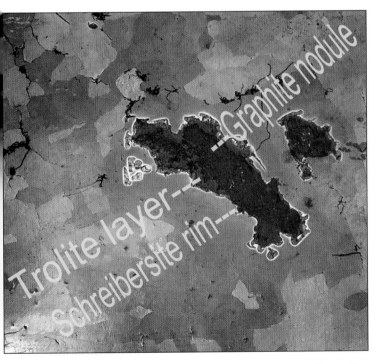

The graphite nodule in the previous photo showing concentric layers of troilite and schriebersite around a graphite interior.

Henbury, Central Australia: These small metallic meteorites are associated with the Henbury Craters; most of these meteorites are small. (Value range F).

Chinga, Siberia, Russia: This is a metallic meteorite that is essentially of the same composition as stainless steel. Because of its high nickel content, these nickel-rich ataxites are very resistant to weathering and rusting. A number of these interesting metallic meteorites have entered the meteorite market, where they have been collected from a strewn field found in spruce forests near Chinga in Siberia, Russia. (Value range D, single specimen).

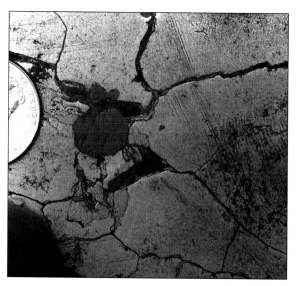

Laddonia(?): Troilite nodule in a metallic meteorite. This large slab was labeled as coming from near the town of Laddonia in northeastern Missouri; however, no meteorite from this region is shown in the "Catalog of Meteorites." It is possibly (probably) a slab of Canyon Diablo, which was purposely misidentified in the nineteenth century and sold as coming from Missouri. It was a common meteorite scam in the nineteenth century to collect meteorites from around what was then known as Coon Butte in Arizona Territory (now Meteor, Winslow or Barringer Crater) and then to fraudulently label them as coming from other places in the US, like this one labeled as coming from Laddonia, Missouri. (Value range C).

Udei Station is an octahedrite containing numerous fragments of stony (silicate) meteorites. Such foreign fragments embedded in an igneous mass (octahedrites and other metallic meteorites are a type of metallic igneous rock) are known as xenoliths (strange rocks). (Value range E).

Miles: Suggestive of a nickel-iron meteorite, but containing some silicate crystals (pyroxene). Meteorites like this, which have more metal than silicate minerals, are still placed as being stony-irons or mesosiderites—the topic of the next chapter. (Value range D).

This entire "Laddonia Missouri" meteorite slab has been set in type metal—another favorite tactic of nineteenth century purveyors of meteorites.

Slice of octahedrite (Find).
Part of a nickle-iron meteorite (find) found near Laddonia, Missouri, Audrain Co. Specimen set in type metal. Pointer indicates a Trolite nodule.

Original label on back of the "Laddonia" slab.

NWA—not indexed: Large numbers of stony meteorites have come onto the meteorite market from the Sahara Desert region of northwest Africa during the first decade of the twenty-first century. Few metallic meteorites on the other hand have come from this region during the same time. Siderites certainly would occur along with the stony's, but they are apparently being "skimmed off" early in meteorites being handled by collectors or dealers. This is one of the few siderites from NWA seen by the author—it came with a small lot of NWA "premiums." (Value range F).

Nuclear Processes and Meteorites

Meteorites record a number of phenomena associated with their parent stars as well as nuclear phenomena associated with stars in general—first discussed is fusion. Stars, like other occurrences of mass in a galaxy, are produced by an accretionary process. The normal sequence of stellar genesis will form a star composed of hydrogen, such **first generation stars** were the first to form following some six billion years after the Big Bang. When a sufficient amount of hydrogen accumulated from a hydrogen nebula, nuclear processes begin in the star's core. In masses similar in size to the sun, the process of hydrogen fusion takes place where hydrogen (protons) fuse with other protons to form deuterium (two protons fused together to form a proton and a neutron). Continued fusion produces three nuclear particles fused together (known as tritium and made up of two neutrons and a proton)—the last step in this process is formation of the element helium (two protons and two neutrons). This fusion of four hydrogen atoms to form helium is an exothermic process (gives off energy); there is a mass loss in the process where the "lost mass" is converted to energy according to the $E = MC^2$ relationship of Albert Einstein. This exothermic fusion process is the energy source of most stars. In a hydrogen star, a multimillion year life time is the rule (in a later generation star, like the sun, stellar life times are measured in billions of years). When most of the star's hydrogen is converted to helium, the star collapses and increased pressure in the star's core initiates helium to undergo fusion. When this happens, the star becomes a nova and elements up to number 26 (iron) are formed—a nova is an exothermic process (that is one that which gives off energy). A nova thus is a stellar explosion that spews out newly formed elements into nearby space and this matter can then become material of a later nebula. Thus, elements higher than hydrogen are now introduced into space, which originally contained only hydrogen and helium. Chondritic meteorites were privy to this process. **Specks of iron found in some chondritic meteorites, as well as the iron coating the armored chondrules, formed directly from iron being thrown off from an exploding star of a nova.**

The other fusion process is a much more intense one. Some stars, if they have sufficient mass, undergo a massive explosion at the end of their "life" known as a supernova. This event, a relatively rare one compared to a nova, is vastly greater in amounts of released energy. Whereas a nova can produce elements up to number 26, a supernova produces elements up to and even above atomic number 92 (uranium)—that is, it can produce elements like plutonium as well as the other trans-uranium elements. Unlike the production of elements up to number 26, the production of these higher atomic weight elements is an endothermic process and the higher the atomic number, **the greater the amount of energy required to produce the element.** A supernova, with its vast energy production, is necessary for the synthesis of the heavy elements!

As in a nova, when a supernova occurs, matter is spewed out into part of the galaxy and this matter becomes available for accretion when it becomes part of a nebula—which it usually does—but in lesser quantities as super novas are much rarer events than are novas.

Matter spewed out by a supernova includes a variety of highly radioactive (unstable) isotopes and elements. These radioactive isotopes, like all radioactive matter, give off heat upon radioactive decay. This heat-of-radioactivity (specifically the heat of short-lived radioactive matter) is what melts the matter, producing the small spheres known as chondrules. Chondrules often have a crystalline structure showing that they crystallized from a molten condition in a zero gravity field. Highly radioactive forms of matter (short-lived isotopes) give off considerable amounts of heat. Radioactive matter also is essentially unstable; it eventually decays into stable decay products. The more unstable an element or isotope is, the greater its radioactivity and the greater amount of heat given off in a short time period from its decay. Highly radioactive forms of matter melted the chondrules, which formed while matter was being spewed out of a nova or a supernova, this heat being produced to a great extent by the decay of a very short-lived, highly radioactive form of aluminum known as aluminum-24. It is also the heat of radioactive decay that modified or metamorphosed many of the chondrules now incorporated into clumped or accreted matter (the other source of metamorphism was produced from high velocity impact). When enough matter clumped together, the heat of radioactive decay caused changes in the chondrules, like the forming of ghost chondrules or if enough matter clumped together sufficient heat becomes available to melt the entire mass. In this way the protoplanets became melted and when the complete mass became molten, **metal originally disseminated through the accreted material** (because of its higher density) settled to the **center of the protoplanet and became its core**—a core surrounded by layers of lower density silicate material. *This melting process thus entirely separated the silicate material from the metallic, allowing elements like those of the platinum group, which have an affinity for iron (the siderophile elements), to become concentrated in the metallic melt forming a core. Other heavy elements like uranium, which has an affinity for silicon, localized where that element is in its highest concentration, thus has uranium been concentrated in the high silicon rocks of the earth's continents and in silicate achondritic meteorites.*

Source of Material in Metallic Meteorites

If the earth is formed of material like that of chondritic meteorites and the chondritic meteorites contain metal, where is all of this metal in earth rocks? It is not found in terrestrial rocks in any quantity compared to amounts found in the average chondritic meteorite. *With the earth, most of this metal (iron and nickel) as well as lesser amounts of the siderophile elements now reside in the earth's core.*

The metal which became the core of either planetoids or protoplanets cooled slowly, forming a distinctive interlocking mass of large crystals which, when a metallic meteorite is cut, gives a distinctive pattern known as Widmanstatten Figures. The cores of planetoids were "opened up" when later impact between these objects shattered them. Thus metallic meteorites with their Widmanstatten Figures represent fragments of the cores of early planetary objects.

Glossary, Chapter Two

Accretion: The clumping of matter together either by one element's affinity for another or by cohesion when the matter was in a hot, slushy condition.

Asteroid: Planetoid-sized objects formed from matter which accreted between Mars and Jupiter. Jupiter's large mass prevented the planetoids from clumping together to form a single planet.

Atomic Number: The number of protons in an atom's nucleus. Elements are designated by this number and are also arranged by atomic number in the periodic table of the elements.

Atomic Weight: The combined number of protons and neutrons in an element's nucleus.

Endothermic Nuclear Process: A nuclear process which takes additional outside energy to continue the reaction. Endothermic nuclear fusion is required to produce elements above number 26 (iron); also, the higher the atomic number, the greater the amount of energy required to produce that element. Elements of high atomic number like thorium and beyond (including the transuranium elements which don't exist anymore {naturally} in our Solar System) are unstable (radioactive) and decay. All transuranium elements in our Solar System have long ago decayed into decay products and some, like element 95 and above, had completely decayed even before the Solar System formed. Radioactive, high-atomic-number elements like uranium-235 and plutonium are utilized as an energy source in nuclear reactors; the energy being released in the reactor being "fossil" energy placed into the uranium and plutonium by the supernova that originally made these elements. Thus, energy coming from a fission nuclear reactor (and whose energy, converted to electricity may be lighting the room as you read this book) is the "fossil" energy given off by a supernova of a long dead star that existed **before** the Solar System existed.

Exothermic Nuclear Process: A nuclear reaction, like fission, that gives off energy. Exothermic fission reactions are the energy source of a nuclear reactor.

Fission: A nuclear reaction involving the "splitting" of high atomic weight atoms (like uranium or plutonium) where energy is released. This energy is also that of a nuclear reactor or an atomic bomb—fission is an exothermic nuclear reaction.

Fusion: A nuclear reaction involving the welding together of atoms as in a sun-type star (which fuses hydrogen to form helium in an exothermic nuclear reaction). Fusion of elements higher than helium is also an exothermic nuclear reaction (up to iron) and takes place in a stellar explosion known as a nova.

Nova: An exothermic stellar explosion in which elements (up to iron) fuse together by nuclear fusion to form elements up to number twenty-six (iron).

Periodic Table: A listing of all elements of the universe according to their atomic number (the number of protons in their nucleus). Hydrogen is atomic number one. Iron is number 26, having 26 protons; elements having up to 128 protons are known, but these do not occur (naturally) today in our Solar System. Uranium is the highest atomic number element to **naturally** occur in our Solar System; plutonium used in reactors and bombs being man-made from uranium.

Planetesimals: Any small, solid object (one to 1,000 kilometers long) that first orbited the sun during the early, formative stages of the Solar System. Planetismals are not large enough to have taken a spherical shape (unless they were molten by the heat of radioactivity). Planetesimals accreted to each other to form the planets.

Planetoid: Asteroid-sized. Many asteroids are probably planetoids or parts of planetoids—others may be broken protoplanets fractured on impact.

Protoplanet: Moon-sized planets, or larger embryos within protoplanetary disks. A protoplanet forms from planetesimals which initially can be kilometer sized. Planetesimals may attract each other gravitationally and then collide. If they contain radioactive materials, it will make their material hot and slushy, which will then stick together rather than fragment, a situation that would not be the case if the material were cold. According to planet formation theory, protoplanets perturb each other's orbits slightly and thus collide in giant impacts to gradually form a real planet.

Radioactive Matter—Short-lived: Radioactive matter is matter which essentially is unstable. It wants to "decay" or revert to a more stable form of matter. There are different levels of this instability. If an element is very unstable (that is highly radioactive), it will have a half life measured in seconds, hours, days or weeks. Such matter will decay and produce decay products in a short time period and since radioactive decay produces heat, such "hot" material will really be physically hot. It is this heat from short-lived radioactive isotopes which was the heat energy present in meteorites and now recorded in them and in their chondrules. Many of the isotopes of low atomic weight elements are the decay products of once highly radioactive, short-lived radioactive matter.

Radioactive Matter—Long-lived: These are the forms of radioactive matter (like Uranium-235, U-238, K-40, thorium, etc.) that still exist naturally in the Solar System. The half lives of these elements and isotopes are measured in hundreds of millions or billions of years, unlike the half lives of short-lived radioisotopes, which are measured in seconds to years. The long lived radioisotopes, like those of uranium, are still around, the short-lived ones on the other hand, having long since decayed to stable decay products.

Siderophile Elements: Elements (such as the platinum group elements or even nickel and cobalt) are elements which are attracted to iron and thus are found primarily concentrated in the earth's core or with other occurrences of metallic iron like that in meteorites.

Supernova: A stellar explosion of a magnitude vastly greater than that of a nova. In a supernova, such a large amount of energy is created that a level of fusion can take place capable of forming all of the elements of the periodic table. A supernova is a much rarer event in a galaxy than is a nova, hence elements above number 26 (iron) generally are rarer than those lower than atomic number 26.

Chapter Three
Stony-Iron Meteorites

Intermediate Space Rocks

These meteorites are intermediate between iron and stony meteorites. They are composed (more or less) of equal amounts of metal and silicates. Three categories of stony-iron meteorites are recognized, the pallasites, the mesosiderites and the lodranites; the latter currently being a rare group known only from two specimens (although this can change in the future with new finds or falls).

Pallasites

Most of the nickel-iron in the chondritic material (which originally formed the earth) now resides in the earth's core. With its relatively high gravity (a consequence of its relatively high mass), the earth's core material is well differentiated from the silicate mantle and crust. In contrast, on small accreted bodies like a planetoid, gravitational attraction is not so great, so immiscible metal and silicate components are not completely separated from each other as they would be with a larger body. This is what the pallasite meteorites are all about, a mix of metal and silicate material which, because of their immiscibility, retain a separation of the two components from each other, but not **completely** separated on a scale necessary to form a separate core and mantle.

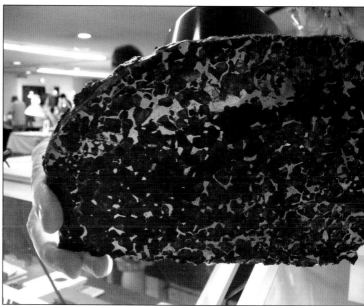

A large back-lit slab of Branin.

Branin, Belarus, Russia: An example of one of these beautiful meteorites in which the olivine crystals have angular rather than round (rain-drop or spherical) shapes. Branin is a pallasite similar to that found at Brenham, Kansas. (Value range A).

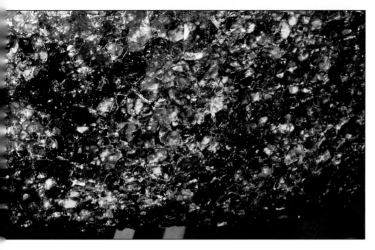

Close-up of Branin slab. *Courtesy of Schooler's Minerals, fossils and meteorites.*

Esquel, Argentina: Esquel is one of the most stable and beautiful of the pallasites. Its olivine also being the yellow-green variety of the gemstone peridot, a semi precious gemstone. This is a small slice of this beautiful meteorite. (Value range F, very small specimen).

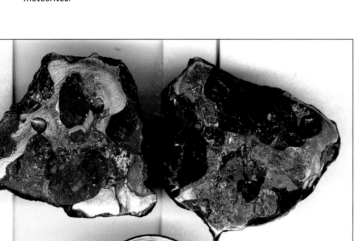

Close-up of backlit slab of Esquel.

Brenham, Kansas: One of the better known and available pallasites, the olivine crystals of Brenham are brownish in color, rather than the yellow-green of many other pallasites. Brenham pallasites show some of the effects of being buried in that they can be prone to oxidize (rust).

Another slice of Brenham.

Purang, China: A large slice of a pallasite, back lit so that the yellow-green olivine crystals stand out. Notice that the olivine crystals are angular and that they are somewhat fractured, the latter being the norm with olivine crystals in pallasites. Rare non-fractured olivine crystals in pallasites (also the gem mineral peridot) sometimes are made into very pricey gem stones and set in jewelry—real gem stones from outer space.

Purang China: Close-up of back lit specimen of a beautiful (and pricey) pallasite. (Value range A).

A large uncut mass of unindexed NWA? Pallasite.

A specimen of an unidentified pallasite, probably from northwest Africa, under reflected light.

Rich Hager with the above chunk.

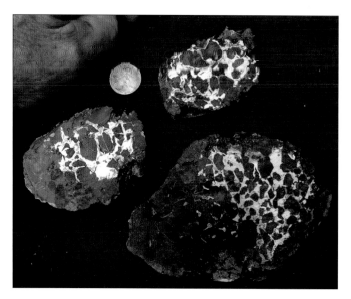

Additional specimens cut from the weathered edge of the above specimen. Sometimes weathered pallasites like this are prone to rusting, as is the case with Brenham. This pallasite appears to be stable and not to rust even on weathered end pieces like this. (Value range D, single slice).

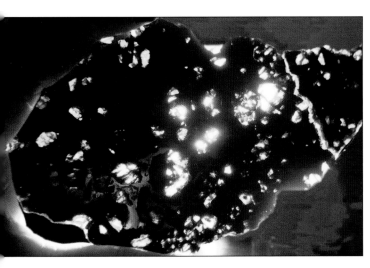

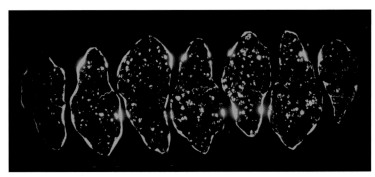

The sliced pallasite specimens backlit.

A backlit slice of the previously shown pallasite after it was sawed and polished. Pallasites are difficult to cut as the metal and silicate components have quite different properties and therefore cut differently. Besides their rarity and attractiveness, this difficulty in cutting them adds to their cost—they are expensive! (Value range A, single slice).

Mesosiderites

Mesosiderites are usually polymict breccias which contain clasts consisting of either silicate material or metal. They represent fragmented material which formed at or near the surface of a planetoid with the fragmentation (as was usual in the early Solar System) produced by colliding objects. Less common are metallic clasts, these having come from one of the colliding bodies, the silicate clasts having come from the other one. Silicate clasts are the most common type of clast in brecciated mesosiderites because silicate material is more easily broken up by impact than is metal.

Some of the slabs cut from the previously shown chunk and mounted in a back lighted display case.

Miles, Australia: Miles is classified as a siderite, however 40% of it is composed of silicate crystals embedded in metal as immiscible components having some of the same signature as the silicate minerals (olivine) in a pallasite. The silicate inclusions in Miles are "blobs" (irregular crystals) of olivine, which have separated from the metal because of their immiscibility with the metal and are not clasts, as is the case with conventional mesosiderites. (value range D).

50

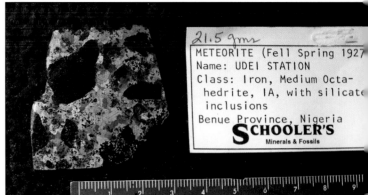

Portales Valley: Flip side of the above slice showing shatter-cone-like structure with metal filled fractures.

Lamont, Kansas: Lamont represents a typical mesosiderite with equal amounts of metal and silicate (black) materials. Note, however, that the top side of this specimen contains no metal! Actual meteorite specimens, like this, can throw a "curved ball" into meteorite taxonomy. (Value range E).

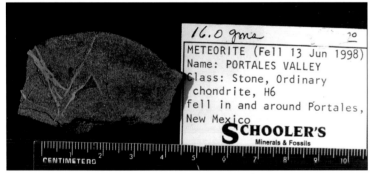

Portales Valley, New Mexico: Portales Valley is considered to be an H6 ordinary chondrite, but it is not ordinary in many ways. It has the many small specks of metal characteristic of an H6 chondrite (which in Chapter six is called a "starry night"); its metal filled cracks especially are **not ordinary**! With some specimens of this 1998 fall, metal fills much wider cracks to the point where the H6, Starry Night components suggest clasts. The cracks were formed by impact, probably before the chondritic material was completely cooled. Impact melted some of the metal or released some of it from a still molten "core" of the parent protoplanet. This liquid metal than filled between cracks formed in the unmelted (but still hot) chondritic material. Note that the flip-side of this slice has a pattern suggestive of shatter cones, another phenomena associated with high velocity impact and impact fragmentation. (Value range E).

Udei Station, Nigeria: Udei Station is classified as a silicated octahedrite (IA). As can be seen in this photo, the large dark silicate components are clasts (some smaller masses may or may not be clasts). Widmanstatten figures (with cubic crystallography) can be seen in the metal; this is one reason that Udei Station is considered an octahedrite.

Vaca Muerta, Chile: Closer to a chondritic meteorite containing a lot of metal. Vaca Muerta is still classified as a mesosiderite. These somewhat-weathered meteorites from the Atacama Desert of Chile have been widely distributed—Vaca Muerta means "dead cow" or murdered cow in Spanish, the word vaccination coming from the same root. (Value range E).

Glossary, Chapter Three

Breccia: In meteorites, a rock or planetary material made up of angular fragments derived from fragmentation on the impact of two objects. The objects may be planetismals, planetoids or protoplanets, or any combination of these. Evidence of such impact is often preserved as a crater on larger objects.

Immiscible: The inability of two materials in a liquid state to mix together, the best known example of two immiscible materials being oil and water. When mixed together, immiscible immediately separate from each other. In meteorites, the two immiscible components are silicates (stone) and iron (metal). This is the reason that metallic core material of the earth has separated from the silicate mantle material and the separation is complete because both materials were once liquid, allowing earth's gravity to allow the denser metal to go to the center, forming the earth's core.

Mesosiderite Meteorites: Meteorites made up of fragments of silicate material that occurs in the form of clasts. These silicate clasts may be set in a matrix of metal or metallic and silicate clasts may both exist. The presence of two distinct types of materials is what gives the two component designation to stony-iron meteorites: stony=silicates, iron=metal.

Planetesimal: A ten to thousand kilometer sized body that contained short-lived radioisotopes (radioactive material), which kept it in a hot and slushy condition. If such a body is later shattered by a high velocity impact after its heat producing radioactive material has decayed, the metallic core can become the source of metallic meteorites. Planetesimals, when hot and slushy, could allow their material to join together producing a protoplanet.

Polymict breccias: Breccias which are made up of clasts of materials derived from different sources and made up of different compositions.

Protoplanet: A sizeable mass of silicate and metallic material which, because of heat produced by short-lived radio-isotopes, becomes molten. When this happens the metal moves to the center of the molten mass forming a core with surrounding layers of silicate material. The crystalline silicate material surrounding the core can become the source of achondritic or even planetary meteorites.

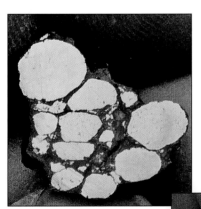

Gujba (also known as Guizba): It's a strange and quite interesting meteorite. (Value range F).

Gujba (Guizba): Another view of this interesting meteorite.

Chapter Four
Stony Meteorites—Finds

Mostly Small Slabs of Small Space Rocks Which Were *Found* by Someone

Most of the specimens illustrated in this and the following chapter are small slices of either finds (this chapter) or falls (Chapter Five). Finds are just that—meteorites which fell, unseen, sometime during the past and later were found by someone. Complete stony meteorites are generally highly desirable as they are rare and therefore are correspondingly expensive when they are available. (An exception to this currently {in 2011} is with NWA meteorites). The shape of some of these meteorites is a consequence of their entry into the earth's atmosphere—such specimens, being informative in matters regarding entry mechanisms and mechanics, are thus intrinsically interesting and correspondingly valuable. The real "gem" of information, however, is in the texture, composition, isotope ratios, and other characteristics associated with events that took place when the material of the meteorite itself was actually assembled—this usually being before or near the beginning of the formation of the Solar System. This is what is emphasized in this and the next chapter and phenomena related to it often can be observed as well in a small slab as in a larger specimen. These small slices, also like larger specimens, are available without the outlay of a large amount of money and thus can allow for a fall or a find (which may not be very large) to be available to many institutions and collectors. These "classic" meteorites also appear to hold their value over time and, thus, might be regarded in some way as a type of investment. Falls may have an appearance that can be more exotic than corresponding finds of the same meteorite type. Sometimes this is subtle, but at other times falls show characteristics and compositions not visible in finds. It might be mentioned, however, that the process of photography, especially digital photography, appears to negate some of the differences between falls and finds.

Some persons, upon their first encounter with stony meteorites, say, "they look like ordinary rocks." The weirdness of metallic meteorites, with their high density, metallic composition, and Widmanstatten Figures are absent in stony meteorites. Yes, superficially they *do* look like "ordinary" rocks, but that is only when they are not critically examined. Stony meteorites almost always contain some metallic nickel iron, a material that is quite rare in terrestrial rocks. Its presence in most stony meteorites is usually conspicuous, but exists in smaller quantities than in the heavier, metallic ones. It is the preponderance of silicates that makes stony meteorites look superficially like "ordinary rocks," but they are far from ordinary! Present in stony meteorites are also the 2-8 mm in diameter spheres known as **chondrules**. Chondrules are found as components in most stony meteorites (although they may not be obvious in all cases) and this also makes them unique as chondrules are an **astronomical phenomena** and **do not occur in terrestrial rocks!**

Matter occurring in large masses, in the form of large rocks or as planetoids over a long period of time and occupying closely spaced orbits, eventually are going to bump into each another. Some of these impacts will be high velocity ones and will produce numerous fragments derived from the original bodies. This is especially the case if the impact took place after most of the original radioactive material in the body had decayed, allowing the mass to lose its slushiness when cooled and thus became really solid. With the presence of radioactive material and its heating effects, impacting masses would be slushy and on impact would stick together much as does wet snow. Non-radioactive-element containing masses in contrast, when they impact, are hard and rigid; when they impact they produce **high temperatures** and form **numerous fragments.** These fragments, in turn, when they later bump into other masses, result in even more fragmentation. If the fragments impact a planetoid or a protoplanet, the result will be the formation of a crater. This is how the extensively cratered terrain seen on the lunar highlands, on parts of Mars, and on the asteroids was formed. With smaller masses, impact produces a sort of churning action (known as gardening)—this effectively redistributes surface material, forming loosely compacted material composed of large numbers of fragments. A geologic material composed of these angular fragments is known as a breccia; individual angular fragments composing a breccia are known as clasts. If this "churned material" is on a planetary surface, it is known as regolith. Thus the moon's surface is covered with such churned material associated with fine fragments (rock flour or lunar dust) which, during the Apollo explorations, often was referred to as lunar soil. This was (and is) a misnomer as **soil is surface material (on the Earth)—a material which contains organic matter (humus)—something obviously not to be found on the Moon.**

Of the three major structures commonly seen in most stony meteorites, two of these are chondrites and clasts. The third are the small masses of metal scattered through a silicate matrix. Impact not only will break material into clasts but can, by the conduction of shock waves, heat and compress loosely cemented material (friable material) into a more coherent form. Weak gravitational attraction also can allow loose material to adhere to a planetoid producing a surface made up of dust and fragments, the planetary regolith. This crushed material (rock flour and clasts), if it becomes lithified, can form a planetoid or protoplanet breccia. The surface of the moon is covered, over large areas, by this sort of material as is the surface of many asteroids. Impact shock can harden this regolith while at the same time producing more fragments—the churning action, referred to as gardening and widespread during the early history of the Solar System is part of the record well represented in many meteorites.

Colony, Oklahoma: A carbonaceous chondrite found in 1975. This type of meteorite is rare as a find as carbonaceous chondrites tend to weather quickly and normally don't survive in our environment, even for a short period of time. This is a type I carbonaceous chondrite, a type with less organic matter than found in the other types (type 2 and 3. Type I's tend to be less delicate so that some, like Colony, can survive as a find. Found 1975, Washita Co., Oklahoma. (Value range F).

Delaware, Logan County, Arkansas: Same specimen as above but with increased contrast. Note that the clasts become more obvious with higher contrast. Note also the weathered rhine at bottom of the slab.

Delaware, Arkansas: L4, ordinary chondrite. Found 1972. Note brecciated texture and what is a large clast at the top of the slab. Such a breccia was produced by impact between asteroids and/ or meteoroids. (Value range F).

El Hammani: Slice of this documented 1980s—collected—NWA meteorite: El Hammani is fresh looking and probably fell in the not too distant past, although the dry desert environment preserves a meteorite much better over a longer time period than ones that fall in areas with humid and wet climates. (Value range F).

Gold Basin, Arizona: These L4 chondritic meteorites have been found in quantity in a stewn field which covers a large part of southern Arizona. Gold Basin meteorites are found by meteorite hunters using metal detectors. Many closely resemble H4 chondrites from northwest Africa. Magnets are stuck to the specimen on the right. (Value range F, single specimen)

Falsey Draw, New Mexico: L6 chondrite. Found 1995. In this meteorite the chondrules are all but gone as a consequence of either thermal or impact metamorphism or both. Specimen under normal contrast. Falsey Draw, Chaves Co., New Mexico. (Value range G).

Gold Basin, Arizona: A sliced example of one of these widely distributed meteorites. Chondrules are observable in Gold Basin but are not as distinct as are those in an L3.

Falsey Draw: Same fragment as above but with higher contrast.

Rusted area

Ghubara, Oman: L5 chondrite. Jiddat al Harasis, Oman. Found 1954. Another example of a documented NWA meteorite. Ghubara is somewhat similar to Farmington (Chap. 5), a meteorite which has been suggested to be part of an Apollo Asteroid. (Value range F).

Gold Basin. Ground terrestralized surface showing metal distribution.

Weathered surface of Gold Basin where metal granules have produced a bumpy surface on desert weathering. (Value range F).

Hebron, Nebraska: H6 chondrite. Hebron, Thayer Co., Nebraska. Found 1965. Note that chondrules are essentially gone; it's an H6—they have been destroyed by thermal metamorphism. The parallel lines are from cutting the slice with a wire saw. (Value range F).

Haxton, Colorado: H4 chrondrite. (Value range F).

Henbury Australia: A characteristically small siderite from the Henbury craters of Australia. (Value range F).

Haxton, Colorado: Same specimen as in previous photo but with greater contrast.

Lake Murray, Oklahoma: Oxide crust from a fossil meteorite. One of a few meteorites that were found imbedded in material of some geological age. Lake Murray (a siderite), when found, had some metal still in it; however, most of the meteorite had become oxide "meteoritic shale" or oxide material like this. Lake Murray was found embedded in an outcrop of Antlers sandstone by the late Allan Graffham in 1932 near Antlers, Oklahoma. Graffham later founded Geological Enterprises, a scientific supply house specializing in fossils. (Value range F).

Lamont, Greenwood Kansas: Found 1940. A mesosiderite with a clear distinction between areas of silicates and metal. (Value range E).

Lunar, (NWA 2995) Feldspathic breccia, Algeria: Found December 2005. A slice of a northwest Africa meteorite found to be a moon rock. This is a porphyritic basalt labeled also as being a breccia. The obvious white crystals of a porphyry are composed of plagioclase. The breccia part is less obvious. Lunars are one of the rarest of meteorites, even though the Moon (Luna) is so close to the Earth. Lunar meteorites, like other planetary meteorites, were ejected from the lunar surface by a process known as spallation, a situation where impact had a sufficient amount of energy for some ejected material to achieve escape velocity. The escape velocity for the moon is low, so this is readily achieved even with medium sized impacts. One wonders why more lunars aren't found—perhaps they will be in the future! (Value range B).

Lemmon, South Dakota: An relative large L4 chondrite slab containing numerous small metal specks which resemble stars in a star studded sky. This distinctive texture is also found in some NWA specimens, which in Chapter Six are referred to as "starry nights." (Value range E).

Lunar (NWA 2995) with label.

NWA 2995
Achondrite, Lunar (ALUN)
Algeria
Found 2005
0.4754 gram

Mt. Egerton, Australia: Find 1941. An aubrite, somewhat stained with the red, iron rich regolith of Australia. Aubrites are composed primarily of enstatite, a magnesium silicate. This specimen consists of a single cleavage fragment of enstatite stained by red clay. (Value range F).

Northwest Africa (2142): A few NWA meteorites have entered the "conventional" meteorite market and have been listed in the catalog of meteorites, as has this one. Most of them, however, remain at large without such indexing. This is a rather unique eucrite breccia. (Value range F, single specimen).

NWA—(not indexed): A large number of the finds of northwest Africa are not indexed in the catalog of meteorites. This meteorite exhibits many conspicuous chondrules. NWA meteorites occur in such abundance that they comprise the subject of Chapter Six. (Value range F).

Northwest Africa 2142: A larger eucrite breccia slab from the same NWA find as above and representative of surface material from a planetary-like body. This slab shows a brecciated texture with an irregular-shaped clast at the upper right made up of interlocking pyroxene and anorthite crystals known as a eutectic. (Value range E).

NWA—(not indexed): Clasts of dark meteoritic material are found inside some NWA meteorites when they are cut and polished. These appear to be clasts of carbonaceous chondrites that came from a carbonaceous object, which collided with a chondritic asteroid forming a regolith breccia. This later became consolidated as a consequence of impact metamorphism. Another more recent impact probably fragmented the specimen and put part of it in its earthly trajectory. (Value range F).

NWA—(not indexed): Another NWA find showing a slightly weathered fusion crust. Some NWA's are finds that fell recently enough for them to retain some fusion crust, like this. A window has been cut to show chondritic composition and chondritic clast embedded in a NWA meteorite.

O'Donnell, Dawson Co., Texas: Found 1942. An H5 chondrite. (Value range G).

Nomenclature of Stony Meteorites

Stony meteorites are distinguished not only by texture (petrologic classification) but also by the name of the place from where they were collected (finds) or were observed to land (falls). For example, Esterville, Iowa, Paragould, Arkansas, and St. Louis, Missouri, are distinctive meteorite types identified by places where they were first seen to fall or were they were found. There is a group of stony meteorite finds that have come on the meteorite market during the past decade which originate from the Sahara Desert, an area which has few landmarks. The place name method of designation is not applicable with these meteorites—a similar situation exists with the bonanza of meteorites found in the blue ice of Antarctica.

Trace Elements in Stony Meteorites

Meteorites differ from Earth rocks, and to a lesser degree from Martian and Lunar meteorites and Apollo Moon rocks by being undifferentiated in terms of many of the elements present in them. Planetary rocks, especially those from the Earth and to a lesser degree those from the Moon and Mars, have their elements well sorted out from each other, certainly more so than do most of the elements found in stony meteorites. Mineral components from planets are also larger in size than are those of most stony meteorites. In earth rocks, minerals can occur as large, distinctive crystals; by contrast minerals present in stony meteorites generally occur as small fragments or as crystals which are difficult to observe with the naked eye. (This is the reason why small pieces of meteorites, like those shown in this chapter, are as adequate to observe their minerals and textures as are larger ones). Complex geologic processes on the earth, acting over long spans of geologic time, have separated various elements from each other, resulting in localized concentrations of different elements like gold, silver, barium, and uranium. In other words, geologic processes on the earth have concentrated elements in what can be the mineable concentrations known as ore. This separation of different elements to form concentrations of them is a major feature characterizing the planets from the smaller heavenly bodies like asteroids. On the smaller bodies, such sorting of different components has either not taken place (as with very primitive meteorites like carbonaceous chondrites) or it has taken place only in an incomplete and partial manner (as with the separation of silicate components from metal in a pallasite). In the primitive meteorites, which includes the chondrites, the rarer elements have not been separated out at all. Chondritic meteorites have most of their rarer elements still mixed with those which make up the majority of an asteroid—elements like **oxygen, silicon, calcium, sodium, aluminum, potassium, magnesium, nickel,** and **iron**. To reiterate, most meteorites, in contrast to earth materials, don't have the rarer elements separated out. Earth materials, on the other hand, have the rarer elements separated from the elements highlighted above. This is also the reason why the elements listed above make up most of the earth's crust (with the exception of nickel). The process of sorting out and separating different elements from each other is known as **differentiation.** A lack of differentiation is a major difference between meteorites and earth rocks. Stony meteorites can even lack a separation between metal and the silicates. In stony-iron, metallic, and achondritic meteorites, some degree of separation has taken place—the first stage toward element differentiation. With the earth this differentiation has taken place to a major degree—otherwise we would not have mineable ore deposits.

Achondrites

A group of stony meteorites exist in which sufficient differentiation has taken place so that metal has been completely separated from the silicates; these are the achondrites. If siderites represent material from a planetoids core, achondrites represent the planetoids mantle and crustal materials. Achondrites represent totally melted and differentiated material in the same manner that igneous rocks of the earth represent totally differentiated material from that of the core, material that originally was derived from chondritic material.

First and Higher Order Distinctions

What is described and dealt with in this work is a general classification of meteorites. With serious meteorite examination, the meteorite is made into what is known as a thin section, which can then be examined under various magnifications as well as being examined by X rays and other forms of short wave electromagnetic radiation. What is presented here, in contrast to this more sophisticated work, is referred to as a **first order examination or macroscopic examination.** This means that the meteorite is examined only with the naked eye, no microscopes, thin sections or more sophisticated observations like microprobe analysis, mass spectroscopy or other methods

capable of extracting information from a meteorite—only visual inspection using the eye. It might be mentioned however that, although microscopic examination has not been used with these images, digital photography has. This is capable of bringing out many macroscopic features which, although visible to the naked eye, are made more striking by digital enhancment.

Fundamental Meteorite Classification

Various meteorite classifications are used in the study of meteorites. The oldest (but easiest to understand) is the tripartite **stony, stony-iron,** and **iron** categories. Also easy to understand and work with is the classification system that uses the name where a particular meteorite was found or was observed to fall. Names like Vaca Muerta, Esterville Iowa, or Brenham pallasite delineate specific types of meteorites recognized and named from their (classic) original find or fall locations. These names bringing to mind a specific and characteristic meteorite type, along with its distinctive signature.

L and H Classifications

H in the nomenclature for chondritic meteorites stands for a high metallic iron content. This refers to **metallic iron** only and **not** to iron **combined with other elements,** as in the silicates. L in chondritic meteorites stands for low metallic iron. An LL chondritic meteorite stands both for low metallic iron (first L) and low ferrous iron combined with silicon in the mineral olivine (second L). Olivine can be composed of two compounds—one known as fayalite, a ferrous silicate, and the other fosterite, a magnesium silicate. With an LL meteorite, the **metallic component** as well as the iron in the **silicate** will be low and the olivine will be fosterite, that form of olivine low in chemically combined iron. H by contrast, refers to high metallic iron, HH refers to both high metallic iron and to high chemically combined iron; in this meteorite the olivine will be fayalite. The number in designations such as **H3** refers to the state of the chondrules—a higher number such as **H5** means that the chondrules are more altered by metamorphism than is the case with an H3. This is supported by the fact that the chondrules of an H5 are not as distinct and observable as those of an H3.

Metamorphism or Change in Chondrules upon High Velocity Impact

Chondrules represent the most elemental (or fundamental) as well as the most ancient major component of stony meteorites, and they are common. Sometimes chondrules are quite obvious; at other times they are less so and on occasions they appear only as "ghosts" of what where originally chondrules. Such a range of variations in distinctiveness of chondrules forms a sequence ranging from clear and obvious chondrules to essentially no chondrules at all—even the chondrule "ghosts" being gone.

This gradational sequence is predominantly an expression of various stages in the alteration of these once clearly discernible meteorite components. It is represented by a numerical series starting with 3 and going to 7. Thus in an H3, the chondrules would be obvious, in an H4 slightly less so, and in an H7 they would be present only as "ghosts," if present at all.

High velocity impact, produced by chunks of material bumping into each other (along with or in addition to heating), changes or metamorphoses the original chondrules. Those subjected to the greatest amount of impact undergo a type of metamorphism (change) with the chondrules become less visible and obvious. If impact is intense enough, the chondrules become obliterated. Impact metamorphism of a similar nature is generated when a large metallic meteorite hits the earth or another large meteor. The crystal pattern represented by Widmanstatten Figures can become partially or totally obliterated by impact. High velocity impact can also soften or even obliterate brecciated textures such as the impact breccias making up asteroidal regoliths. As with chondrites, various stages of distinctiveness of brecciated textures form a sequence ranging from a very clear breccia to a condition where only ghosts of breccia fragments exist. The distinctiveness of breccia clasts forming again a "spectrum of variations" between the two extremes of **very obvious** to **vague clasts** in the same manner as with chondrules.

Terrestrial Weathering of Stony Meteorites

Meteorite finds with a long residence time on the earth's surface react with it and stabilize somewhat. This stabilization process consists, in part, of oxidation and hydration of iron silicates, producing a secondary silicate and hydrated iron oxide (limonite), which adds a tan or brownish color to the meteorite. Fresh stony meteorites can contain compounds that basically are unstable when introduced to the earth's atmosphere and its water; these compounds, in other words, don't like our environment and with long earthly residence times they chemically change and come into equilibrium with it. The meteorites in this chapter, being finds—some of which having had long earthly residence times—have come into equilibrium. In contrast, those collected only a few years after they fell may retain some of the freshness of falls, the topic of the following chapter.

Carbonaceous Chondrites

These primitive, fragile, and peculiar meteorites are rarely encountered as finds. One exception is in some northwest African (NWA) meteorites where carbonaceous chondrite clasts occur when they were embedded in an asteroidal regolith. This regolith was then metamorphosed by impact and the carbonaceous chondrite (cc) fragments protected by being embedded in more normal meteoritic material. Carbonaceous chondrites as falls contain a complex of organic material. As finds, this organic material, as well as clay minerals, are both less obvious and significant; organic mater may also be introduced as an earthly contaminant and the clay minerals (unique in fresh falls of cc's) may form from terrestrial weathering of silicates.

Potter, Nebraska: Brecciated, L6 chondrite. Found 1941. Clasts are hard to see because of both a long earthly residence time and impact metamorphism, which blurred the clast boundary sometime before their entry onto the earth. Potter, Cheyenne Co., western Nebraska. (Value range F).

Powellsville, Ohio: An H5 chondrite found 1990. Powellsville, Scioto Co., Ohio. (Value range F).

Seibert, Colorado: An L6 chondrite. Seibert, Kit Carson Co., Colorado. (Value range F).

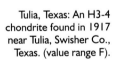

Selma, Alabama: Found 1906. An H4 chondrite in which the chondrules in this slab show up fairly well. The specimen is broken along a limonite stained fracture. Dallas Co., Alabama. (Value range F).

Close-up of Seibert showing chondrules and weathered rind at the left.

Tulia, Texas: An H3-4 chondrite found in 1917 near Tulia, Swisher Co., Texas. (value range F).

Tulia, Texas: Another specimen from the same find as shown in previous photo. (Value range G).

Vyatka. Same specimen—darker and higher contrast.

Vyatka, Russia: An H4 chondrite found in 1992 on the shore of the Vyatka River near Sovtesk, Kirov region, Russia. This is an example of a chondritic meteorite with many small specks of metal. The author has referred to this texture, which is also found in some NWA meteorites, as a "starry night." (See Chapter 6.) (Value range E).

Vaca Muerta (Dead cow): A type of (somewhat ugly) mesosiderite which has been widely distributed among collectors. (Value range F).

Glossary, Chapter 4

Lithification (lithified, adj.): The process by which loosely compacted material is cemented or compressed together. Loose, fragmentary material on a surface in space can be lithified from shock waves produced on impact. Lithification can also come from partial melting (softening) or from the heat of radioactive element decay. Lithification in meteorites is related to metamorphism.

Metamorphism: Literally "a change in form" of something, including rocks and meteorites. Metamorphism in space rocks (meteors) can take place by heating (thermal metamorphism) and by impact (impact metamorphism). Change in a space rock (now a meteorite), when it is on the earth's surface also takes place—this is known as weathering, or terrestrialization, and is not considered metamorphism.

Rock flour: Pulverized rock or rock dust produced from high velocity impact. Sometimes the formation of a breccia by high velocity impact can be distinguished from breccia that originated from another process by the presence of considerable amounts of rock flour occurring between breccia clasts.

Chapter Five
Falls

Consisting mostly of small slices of usually small space rocks that were actually observed to fall and shortly thereafter were collected.

The illustrated specimens in this chapter are small slices of meteorites that actually were observed to fall from space. Meteorites collected as falls are generally the most desirable; they also are rare and therefore are correspondingly expensive when they are available. Meteorites obtained as falls are particularly valuable to science, as they can contain components quickly lost once the meteorite encounters our atmosphere or the meteorite's gaseous components dissipate. On this matter, it is interesting to note that witnesses of the Murchison, Australia, fall in 1969, a carbonaceous chondrite (cc), reported solvent-like odors associated with it just after it was collected. These vapors were low molecular organic compounds contained within the meteorite that were quickly escaping. Falls are the "freshest" meteorites available and may contain components and related information that otherwise would be lost in finds, even relatively fresh ones. Fresh falls may also contain chemical compounds that react quickly to atmospheric oxygen and/or water vapor and therefore are not present in finds.

Chondritic Meteorites & "Weird" Chemistry: Phenomena Especially Characteristic of Falls

Having been formed from a nebula in space, chondritic meteorites might be expected to contain some unusual components compared to terrestrial rocks, and indeed they do! This is especially the case with meteorite falls, which, unlike most finds, **have not interacted with the earth's atmosphere**.

Falls can retain the "weird" components of deep space that don't survive in the earth's atmosphere (or don't survive very well). One group of primitive meteorites known as the carbonaceous chondrites can contain unusual minerals beside the (inorganically formed) organic compounds for which they are famous. In these meteorites, as well as in some achondritic meteorites, can be found chemical compounds totally absent in terrestrial rocks. Some of these include various phosphides, chlorides, sulfides (including the mineral oldhamite—calcium sulfide, the compound used to put the glow in "glow in the dark" plastic), hydrides, and carbides of iron and magnesium. Magnesium sulfide, and

Allende, Mexico: Fell Feb. 8, 1969. Type III carbonaceous chondrite (CV3). Massive detonations accompanied by fireballs and other brilliant displays heralded Allende—the largest fall of a carbonaceous chondrite known. Chondrules show up especially well in Allende, in part because, as with all carbonaceous chondrites, they have never been heated to any degree. As such, they are regarded as the most primitive type of meteorites—original stuff—(inorganically formed)—organic compounds and all. (Value range D).

calcium carbide have also been found—but only in very fresh falls as these compounds react vigorously with water (as well as with water vapor in the atmosphere). Calcium and magnesium phosphides have been reported as well as have other phosphides. The iron phosphide schribersite is another exotic mineral which frequently forms as a halo around trolite nodules in nickel-iron meteorites; however, it, unlike most other phosphides, is stable in the earth's atmosphere. Ferrous chloride (lawrencite) is one of the most "pesky" of the exotic compounds found in meteorites. Lawrencite reacts with water vapor of the earth's atmosphere forming hydrogen chloride and ferrous hydroxide, the latter being ugly, rusty looking stuff that can ooze from a freshly cut meteorite's surface. Lawrencite is that extraterrestrial mineral of greatest bane to meteorite collectors.

These exotic compounds, with some exceptions like lawrencite, are usually noticeable **only with fresh falls**.

Stony meteorites, with a long residence on the earth, like the NWA's, have had time to slowly react with atmospheric components and to have reached chemical equilibrium with both terrestrial water vapor and oxygen. Sulfides such as oldhamite (calcium sulfide) will oxidize to calcium sulfate, which is gypsum (a hydrous calcium sulfate), carbides such as calcium carbide become calcium carbonate (calcite), and hydrides become hydrates by combining with terrestrial oxygen; when this happens these minerals are then stable. Chondritic meteorites occurring as finds also come to more closely resemble terrestrial rocks and lose some of their "chemical strangeness," which usually includes the exotic components. **However, a long earthly residence time for stony's doesn't change them too much in appearance from recent falls as the strange or exotic components are generally minor components compared to the predominant nickel-iron and silicate minerals.**

Allende: Small fragmented specimen with normal contrast. Note that the chondrules are packed close together. 1969 was a banner year for space related phenomena—the Apollo moon landing, the Allende meteorite bonanza, and the Murchison fall. (Value range E).

Allende: Same specimen but with higher contrast, which brings out the chondrules. Note large, yellowish chondrule at the center.

Allende: Slab showing various sized chondrules; chondrules show up especially well in carbonaceous chondrites as they have never been changed by heating, a process which would not only destroy the organic components but which also would alter the chondrules.

Allende: Same slab, but with more contrast. Chondrites generally show up better in falls than in finds, which (in some cases) have weathered on the earth's surface.

Allende: Same slab as above but in more subdued light.

Achondrites

These meteorites are essentially igneous rocks—as with the cc's, most are known as falls. Igneous rock is rock that forms as a consequence of the cooling and crystallization of molten silicate material known as magma. With the genesis of these meteorites sufficient quantities of chondritic material becomes clumped together so that the heat of decay of radioactive elements present in the clump are able to melt the entire mass, not just to modify or "cook or metamorphose it" as would be the case with a smaller clump. With a large mass of metallic nickel-iron present in the melted mass going to the center to produce a core, an outer layer of molten silicate material surrounding this core makes up a mantle. Achondrites represent fragments of this mantle material, which, when it cooled and crystallized, formed a mass of interlocking silicate-mineral crystals.

Aubrites

These are achondrites composed primarily of the mineral enstatite, which is a pyroxene. Enstatite is an aluminum magnesium silicate which, as it contains no iron, is light colored (usually light grey). From spectral observations of asteroids, the aubrites may originate from asteroid 44 Nysa, which is a near earth asteroid.

HED Group (Howardites, Eucrites, & Diogenites)

This group of achondritic meteorites has been suggested to originate from Vesta, the third largest asteroid in the Solar System. Howardites may represent the regolith breccia of Vesta and eucrites may represent its crustal basalts—they being much the same as terrestrial basalts. Diogenites are made up of pyroxenes, which may have originated from deep within Vesta's crust or may have originated from part of its mantle; they can be made up of beautiful green crystals of magnesium pyroxene.

66

Bensour, Morocco: Fell Feb. 10th, 2002 near the Morrocan-Algerian border. LL6 chondrite. (Value range F).

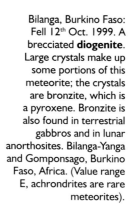

Bilanga, Burkino Faso: Fell 12th Oct. 1999. A brecciated **diogenite**. Large crystals make up some portions of this meteorite; the crystals are bronzite, which is a pyroxene. Bronzite is also found in terrestrial gabbros and in lunar anorthosites. Bilanga-Yanga and Gomponsago, Burkino Faso, Africa. (Value range E, achrondrites are rare meteorites).

Bilanga. Fell Oct 12, 1999. Opposite side of the specimen shown in the previous photo. Note small fragments of the specimen at the upper left. Diogenites are quite friable, these small fragments fell onto the surface of a flat bed scanner when the image was being made.

Berduc, Argentina: Fell April 8, 2008. Brecciated (and cracked). The fractured condition of this meteorite is especially conspicuous. This meteorite is similar to Peekskill, New York. (Value range F).

Cold Bokkeveld, South Africa: Fell Oct. 13, 1838. A type three carbonaceous chondrite. Pieces of this meteorite resemble a piece of dull coal or carbonaceous shale, an earth rock often associated with coal beds. Cold Bokkeveld, Cape Province, South Africa.

Cold Bokkeveld, Cape Province: Enlarged view of small fragment of this coal-black, carbonaceous chondrite.

Farmington, close-up: Chondrules are not evident; they have been destroyed by thermal metamorphism.

Farmington, Kansas: Fell June 25th, 1890. Olivine-hypersthene chondrite. Chondrules are **not evident.** The meteorite resembles a porphyritic basalt from intense heating, which came almost to the melting point of the original chondritic mass. Farmington is believed (possibly) to be part of an Apollo Asteroid. (Value range E, single slab).

Gao-Guenie (Upper Volta): Fall, March 5, 1960. Olivine-bronzite chondrite. A large number of these dark stones with well developed fusion crusts were collected by and brought onto the meteorite market by "The Meteorite Man," Robert A. Haag of Tucson, Arizona. Most of these specimens exhibit a black fusion crust, as shown here. (Value range E).

Gujba (also known as Guizba), Nigeria: Fell April 3, 1984. A unique meteorite containing metal spheres. These are not chondrules; Gujba somewhat resembles a Bencubbin stony-iron but is otherwise a unique and peculiar meteorite. What's especially peculiar about it are the spherical masses of metal—sometimes described as metallic chondrules. These metal "nodules" are unique, but are incorrectly referred to as being chondrules. (Value range E, rare meteorite type).

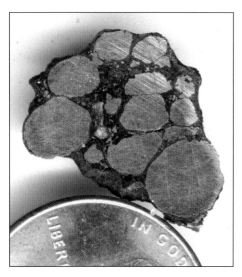

Gujba: Close-up view of this peculiar (and unique) meteorite, which may have a metal sheath but otherwise is composed of silicates.

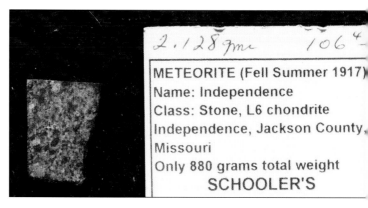

Independence: Fell Summer 1917. An L6 chondrite. Independence, Jackson Co., Missouri. (Kansas City area). (Value range E).

Independence: Close-up of the previously shown specimen.

Holbrook, Navajo Co., Arizona: Fell July 19, 1912. L6 chondrite. Small slab from a particularly large specimen of this fall of predominantly small, stony meteorites. What appears to be a large mass of the iron sulfide mineral troilite is present in this slab. (Value range E, a large specimen for Holbrook and the trolite inclusion is unusual for its size in a stony meteorite).

Independence: Another slab of this meteorite from the Kansas City area shown with normal contrast.

Independence: Same slab, but with more contrast. Note that chondrules are absent; metamorphism has obliterated them.

Jumapalo: Back side of previously shown small slab.

Jumapalo, Java, Indonesia: Fell March 13, 1984. A L6-ordinary chondrite. This specimen is friable, as are many stony meteorites associated with falls. It also has an ashy grey appearance—a characteristic of many falls. Some oxidization has taken place even with this fall. Note metal mass in a second clast to the right. Karanganyar district, central Java.

Kilabo, Nigeria. Fall July 21, 2002. A brecciated chondrite very low in iron. Kilabo resembles Peekskill, it has the same sort of brecciation and texture as Peekskill—possibly both originated from the same asteroid. (Value range F).

Kilabo. Another view—brecciation is very evident.

Kunashak, Russia: Fell June 11, 1949. An L6 chondrite. Noticeable brecciation with conspicuous cracks, referred to as "veins." Kunashak, Chelyabinsk region, Russia. (Value range D, large slab).

Millbillillie, Australia: Fell Oct. 1960. Achondrite- Ca rich **eucrite, a crystalline igneous rock meteorite**. This meteorite is characterized by being composed primarily of **white calcium plagioclase** (anorthite) into which are set black or grey pyroxene crystals. (Value range E).

Kunashak. Same specimen as above, but taken in low angle reflected light to emphasize metal components.

Murchison. Australia: Fell Sept. 28, 1969. Type I (CM2). Note the small chondrules, a characteristic of many carbonaceous chondrites. This small slab has been cut with a wire saw—the vertical lines are from this. Cutting with a fine wire saw loses less of the meteorite to dust than with a circular rock saw blade, even a thin one. Murchison is one of the most unique and well studied of carbonaceous chondrites. Its fall, along with the Apollo moon landing, was one of the "extraterrestrial highlights" of 1969. (Value range E).

Kunashak: Same specimen as above but taken in direct light from above to de-emphasize metal.

Murchison. Group of small pieces. Note again the small, but distinct chondrules, a characteristic of type I carbonaceous chondrites.

Murray, Calloway Co., Kentucky: Fell Sept. 20, 1950. This carbonaceous chondrite was recovered after a spectacular fireball was seen on Sept. 20, 1950. A number of organic compounds were reported from Murray, including amino acids. Their presence in meteorites was still questioned at the time as it was believed by some scientists to be a contaminant until verified by Murchison, which has an overall higher percentage of organics.

Murchison. This meteorite has some of the highest concentration of extraterrestrial organic material known in meteorites. A considerable range of compounds have been identified in Murchison, including high molecular weight hydrocarbons, ketones, aldehydes, and even amino acids. Some have reported over 90 different organic compounds in Murchison including some not known from biogenic sources. (Recent high tech analysis identified even more organic compounds.) These inorganically produced organic compounds are believed to have been produced in a nebula at low temperatures and to have never been heated—otherwise the organic components would have been destroyed.

Nakhla, Alexandria, Egypt: Fell June 28, 1911. Nakhla is a Martian meteorite, an SNC. The "N" in SNC stands for Nakhla. (Value range A).

Murchison. Another small piece. (Value range G).

Nuevo Mercurio, Ranacho Santa Cruz, Zacatecas, Mexico: Fell Dec. 15, 1978. Slice of a small stone with fusion crust. Oxidization of an iron bearing component has taken place. Brown material is ferric hydroxide. (Value range F).

Norton, Kansas: Fell Feb. 18th, 1948. Single crystal from a relatively large fall, especially considering that this is a rare type of meteorite, an **aubrite**. Norton is composed mostly of enstatite, a magnesium silicate. This is a nice single crystal of that mineral. (Value range G).

Norton: A group of small fragments of Norton (note penny for scale). Dark areas are small specks of iron, which, since its fall in 1948, have oxidized (rusted). (Value range E for group).

Norton: Larger fragment than those shown above, where the specimen has been placed on a flat bed scanner. Note small fragments at 10:00 that have fallen from the specimen. Aubrites, like diogenites, are crumbly and friable. (Value range E).

Ochansk, Perm Region, Russia: Fell Aug. 30, 1887. H4 chondrite. Relatively friable is this fresh meteorite. (Value range F).

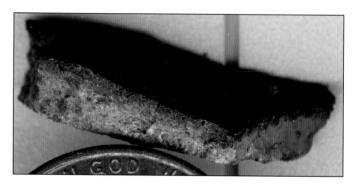

Ochansk, Russia. Shows fusion crust on the outside of the slice shown above.

Ourique, Beja district, Portugal: Fell Dec. 28, 1998. H4 chondrite (brecciated?).

Ourique: Note the high amount of hydrated iron oxide (rust) in this specimen, even though it is from a fall. Such oxidization is usually not the case with falls, but here some component in the meteorite reacted with either oxygen or water (or both) producing this brownish mineral. Many meteorites don't like our atmosphere and in this case this fall reacted with it quite quickly. (Value range F).

Peekskill: The fall of this meteorite was a nice event overall. No one got hurt, many persons in the eastern part of the country got to see a nice fire ball (often during fall football games), and the smashed Chevy Malibu, originally thought to be the work of vandals, was purchased at a premium price by a meteorite dealer. Some nice H6 breccias also came with the deal— a win-win situation for all.

Portales Valley: Fell June 13, 1998. An H6 chondrite. Shows impact related iron filled cracks. These iron filled cracks formed from impact some time in the early history of the Solar System—a time when radioactive material, capable of producing sufficient heat, melted the iron, allowing it to fill impact formed cracks. (Value range E).

Peekskill, New York: Fell Oct. 9, 1992. H6 (brecciated) chondrite. The spectacular fireball created by this meteorite when it entered the earth's atmosphere was well-documented by many sources, including video cameras at fall football games. The meteor landed in Peekskill, New York, where it hit a Chevy Malibu in the right rear. Meteorites that hit something like the Malibu are referred to as "hammers."
Peekskill has a distinctive brecciated texture—it retains an especially fresh appearance characteristic of many falls. Peekskill is similar to Kilabo—possibly both being from the same parent asteroid. (Value range D, high value as a hammer is quite desirable with some collectors).

Portales Valley: Flip side of the previous shown slab, showing a crack pattern suggestive of and often associated with shatter cones. Photo has been taken at a low angle to emphasize the metal filled cracks.

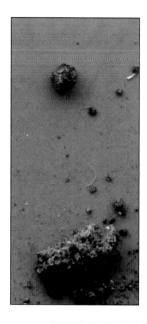

Saratov, Russia: Fell Sept. 6, 1918. L4 chondrite. This meteorite is quite friable, with the chondrules separating readily from the matrix. Fell near Donquz, Penza, Saratov region, Russia. (Value range E).

Tagish Lake, British Columbia, Canada: Fell Jan. 18, 2000. Small chondrules are a characteristic of many carbonaceous chondrites (cc's). Type I carbonaceous chondrites are rare meteorites and are also quite fragile. On entry into the earth's atmosphere they generally are destroyed by friction and ablation. Because of their softness and composition, cc's can produce spectacular fireballs, as was the case with Tagish Lake. Small fragments like this are what is generally available on the meteorite market of these rare meteorites. Close up photo of Tagish Lake fragments. (Value range F).

Sikhote-Alin, Siberia: From a massive fall that took place Feb. 12, 1947, in eastern Siberia. The Sikhote-Alin meteorite must have been a loosely bound swarm of shrapnel-like chunks like those shown here. Vast numbers of these shrapnel-like meteorites pelted the ground, producing small impact craters and holes as well as removing both branches and needles of coniferous trees in a heavily forested area. This was one of the largest falls of metallic meteorites known. (Value range F, a single specimen).

Sikhote-Alin stamp: The fall of the vast number of metallic meteorites in 1947 over the eastern part of the Soviet Union produced the spectacular display represented on this postage stamp. The large strewn field of Sikhote-Alin was not utilized as a meteorite resource until after the end of the Soviet Union, when numerous examples (which resemble shrapnel) of this large fall became available on the meteorite market. The locking-up of an educational resource like Sikhote-Alin, with its thousands of specimens, by not allowing any of it to be collected or distributed (as was the policy of the Soviet Union), takes away from citizens what is an excellent hands-on educational resource as well as a source for local monetary gain for the region where the meteorite fell. In some ways, a meteorite fall like this is analogous to money falling from the sky—a nice proposition, especially if no one gets hurt or killed. But by locking up such a natural resource as this, it is not being preserved, it is being locked-up possibly to be destroyed by weathering.

Tagish Lake: This meteorite landed on the frozen surface of Tagish Lake. The cold temperatures and careful collection techniques enabled volatile organic compounds present in this meteorite to be identified. These compounds include aromatic hydrocarbons, ketones, and aldehydes. Such compounds had to have formed in a cold environment and to have been kept at a low temperature—the temperature of deep space—to survive. To preserve a carbonaceous chondrite requires the absence of any heat generating, radioactive material in its environment after it formed in a pre-solar nebula. (Value range F).

Tamdakht, Ouarzazate Morocco: Fell Dec. 20, 2008. Some oxidization of iron has taken place on this meteorite since entry. Note the perfectly round chondrules. (Value range F).

Tamdakht: Same photo as above but with chondrules noted.

Tatahouine: Typical fragments of this unique meteorite under normal lighting. These are small crystals of the mineral hypersthene, a magnesium pyroxene (orthopyroxene). (Value range F for group).

Tatahouine, Tunisia: Fall June 27th, 1931. An achondrite shower fell at Tatahouine, Tunisia, in which thousands of fragments like this fell as a shower dispersed over a small strewn field. The fragments were composed of lovely green orthopyroxene crystals (hypersthene) with accessory chromite. The meteorite was a fragmented **diogenite**—it fragmented upon entry into the atmosphere—the small pieces slowing down sufficiently so as not to undergo ablation and vaporization. Small specimens can still be collected in the strewn field by sifting the soil, which yields these small fragments, although those found now show the effects of terrestrial weathering. (Value range F for group).

Tatahouine: Same fragments as shown above but under more intense lighting and contrast.

Tatahouine: Specimens lit at a low angle. Note translucent hypersthene crystals. Hypersthene is a magnesium pyroxene which is found in mafic igneous rocks of terrestrial origin like gabbro.

Warrenton, Missouri: Fell Jan 3, 1888. The placing of this small piece of carbonaceous chondrite on a flat bed scanner to make this scan, crushed it—carbonaceous chondrites are fragile and very friable. It would be nice if there was more of it! (Value range G, before being crushed).

Udei Station, Nigeria: Fell Spring 1927. Medium octahedrite with silicate inclusions. (Value range E).

Weston, Connecticut: Fell Dec. 14, 1807. Thomas Jefferson is said to have stated, of circumstances surrounding the fall of this meteorite in 1807, "I would be more inclined to believe that a Yankee professor would lie [Benjamin Sillman] than that a stone would fall from the sky." Contrary to this pronouncement by Jefferson, the observed fall and its documentation gave scientific credence to the idea that space rocks could indeed fall from the sky. Science prior to the nineteenth century was skeptical regarding meteorites being what they really are, rocks from space! (Value range D, a historically significance meteorite).

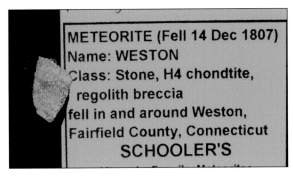

METEORITE (Fell 14 Dec 1807)
Name: WESTON
Class: Stone, H4 chondtite,
regolith breccia
fell in and around Weston,
Fairfield County, Connecticut
SCHOOLER'S

Weston with label.

Zag: Enlarged to show oxidization of iron. There is a similarity with Zag and Weston, both of which have an almost identical texture and composition. Did they come from the same asteroid?

Weston: Notice the dark clast near the bottom.

Zag: Fell Aug. 5, 1998, in the vicinity of Zag, east of Tan-Tan Morocco. H5 chondrite. (Value range G).

Small meteorites have been found embedded in the roof of "big box" buildings like these. Small meteorite falls sometimes overtake the earth in the same direction it is traveling in its solar orbit so that their entry velocity is low enough that they do not produce a noticeable fireball. Such a fall may go unnoticed unless the meteorite is actually found or if it hits and damages some work of man like these warehouses. A few instances have been reported where roof damage from a small meteorite caused a leak with the meteorite still embedded in the roof—in one of these the leak was above the lingerie department of a big box store.

The Paragould Meteorite That Wasn't!

The fall of a large stony meteorite took place near the northeastern Arkansas town of Paragould on November 8, 1936. Spectacular aerial phenomena accompanied this fall. The fireball created on its entrance into the earth's atmosphere lit up the sky and was seen over a large portion of the Midwestern United States. Following some of the same techniques used by Harvey H. Ninninger, the first to assemble a large collection of "new" meteorite falls and finds in North America, the author made up a series of "reward for meteorites" posters, which he placed in country stores of northeastern Arkansas, a region primarily of flat terrain devoid of rocks, so that any rocks encountered associated with agricultural pursuits had a reasonable chance of being meteorites. One response to these posters appeared especially promising. A farmer near Paragould, remembering the 1936 fall, had found a dark, heavy stone the day after the Paragould fall. He had kept this rock for decades, but a few years previous to my inquiry had given it to a Girl Scout camp where he sometimes worked. In exchange for this "meteorite," I offered the camp a nice collection of Missouri fossils—feeling with certainty that this was probably a part of the Paragould fall. He thought that the offer would be accepted and over the Christmas holiday would be at the camp and would bring back the stone in exchange for the fossils. A return trip to Paragould, shortly after the new year found the farmer with the stone. It was dark, heavy, and somewhat metallic looking. It looked meteorite-like but different from descriptions of the Paragould meteorite and not quite like the specimen I had seen previously in the Field Museum of Natural History in Chicago, which I had visited with my parents a few years earlier.

There was a particular problem with the "meteorite," however. On rubbing it on a hard, white tile it left a red streak—the reddish color typical of hematite, **which is ferric iron oxide**. Meteorites, particularly unweathered ones like a fall, don't contain **ferric oxide**, that form of iron in which maximum oxidization has taken place. (Meteorites will contain either ferrous oxide or metallic iron, but almost never this highly oxidized, third form of iron). On closer examination, the heavy stone looked like a chunk of Missouri hematite, specifically an ore sample from an iron mine that produced large amounts of hematite at Iron Mountain, Missouri. The "meteorite" turned out to be a meteor-wrong made of hematite. Apparently after the Paragould fall, the farmer (and many other persons in northeastern Arkansas) was meteorite conscious and noticed dark, heavy rocks, which otherwise would have been ignored. The fact that there were railroad tracks nearby where the stone was picked up might explain the hematite from Iron Mountain—it fell from a car carrying an ore shipment on the Missouri Pacific Line in Paragould, a line which also passed through Iron Mountain, Missouri. It was a nice try but reinforcing the fact that most "meteorite" occurrences, like this, turn out to be false alarms.

Carbonaceous Chondrites (cc's)

One of the most peculiar and unique meteorites—which because of their fragility are almost all known from falls—are the carbonaceous chondrites. They are unique because they indicate the presence of water (from the presence in them of hydrated silicates {clay-like minerals}) but even more unique in that they contain a considerable variety of **organic compounds**. These organic compounds include ketones, aldehydes, and alcohols, as well as other solvent-like compounds. Probably the most intriguing organics found in them are a variety of amino acids, nitrogen containing compounds that are the building blocks of proteins and life. These compounds have been identified in the falls of Murchison, Orgueil, and Tagish Lake, as well as in other falls where they can constitute up to five percent of the meteorite's mass. As falls, cc's can also contain clay-like minerals, minerals containing chemically bonded water—a chemical compound absent from other fresh meteorite falls. Carbonaceous chondrites are also what is considered to be the most primitive of all meteorites. Primitive material derived from different parts of a nebula—the silicate portion representing star related, high temperatures, the organic and hydrated portions representing low temperature materials formed in a nebula. Awareness of these unique and significant meteorites first took place in the 1860s when the first one to be recognized came from a fall in France, the Orgueil meteorite.

Allende, Mexico: This is one of the most readily available carbonaceous chondrites, large amounts of it having been collected after its fall on Feb. 8, 1969. The dull, uninteresting(?), appearance of Allende is (partially) a consequence of its clay mineral content. It does contain beautiful chondrites however. (Value range F, small 30 gram piece).

A group of carbonaceous chondrites containing a maximum amount of organic matter are the CV types (in older literature known as type III cc's). Looking like pieces of a charcoal briquette, cc's are probably one of the least attractive of any meteorites, but considering their organic matter arguably they may be the *most* interesting. Associated with cc's also can be found highly reduced compounds like carbides, specifically calcium, magnesium, and iron carbide. With regard to the organic compounds in cc's, calcium carbide is especially noteworthy. This substance, industrially known as "carbide," when it comes in contact with water, produces a vigorous chemical reaction producing acetylene, an inorganically formed, reactive **organic compound**. The industrial production of acetylene, a compound forming the basis for the synthesis of many industrial polymers like plastics and rubber, starts with calcium carbide. Indeed, a major industrial player in the utilization of this chemistry is Union Carbide Company, a corporation which capitalized upon the fact that calcium carbide could be produced in large quantities from the DC (direct current) dynamos that came from the late nineteenth century electrical industry fostered by Thomas Edison. The part played by natural carbides in the formation of the organic compounds found in cc's still remains unknown but the fact that acetylene, an (inorganic) organic compound, can be produced from calcium carbide and water might suggest some possible chemical pathways.

Because organic compounds can be produced by inorganic means (and organic compounds have been found in the tails of comets), cc's are believed by many meteorists and astronomers to be the spent, solid material of comets; cc's also are fragile meteorites. As a consequence of their containing organic compounds, cc's—since their recognition in the 1860s, have been a target for that most elusive goal of science, proof of the existence of life somewhere else in the universe. Investigations of meteorites in the hope that they might contain fossils first took place in the mid-nineteenth century. Advances in the 1950s in both analytical chemical techniques and the perfection and widespread availability and use of the electron microscope gave impetus to more focused work on cc's.

Advanced analytical techniques proved beyond a doubt that the wealth of organics reported in cc's were really real and were not from contamination. Ever since organics had been found in the Orgueil, France, fall of 1864 there had always been some question as to their actually being a part of the meteorite rather than contaminants from the earth and handling. The fall and immediate recovery of another carbonaceous chondrite in Kentucky, the Murray meteorite (fell September 20, 1950), provided fresh material for more sophisticated analysis that, as with Orgueil, showed the presence in the meteorite itself of a variety of often complex, organic compounds.

Colony, Oklahoma: One of the few carbonaceous chondrites to be collected as a **find** rather than as a **fall**.

Murchison, Australia: Fell Sept. 28, 1969. One of the most carbonaceous looking of the carbonaceous chondrites, Murchison also fell in 1969, a banner year for space rocks. Hundreds (some report thousands) of different organic compounds have been identified from Murchison, including 90 different amino acids. Murchison is a CM2, formerly called a type III carbonaceous chondrite.

Murray, Kentucky: Type III (CM2), fell Sept. 20, 1950, east of Murray, Kentucky, near Kentucky Lake. Murray was analyzed by modern means (gas chromatography) for its organic compounds. Amino acids were tentatively identified; however, these were suspect as having come from contamination. The 1969 fall of Murchison confirmed the amino acid identifications in carbonaceous chondrites.

Unindexed NWA: Brecciated meteorites occurring in lots of NWA's occasionally show clasts when cut. Some of the clasts are dark and may be clasts of carbonaceous chondrites. Breccias such as these may be from an asteroidal regolith, which was "gardened" by impact when some of the impacting objects, being carbonaceous chondrites, may have come from comets. Unlike cc's on the earth, in space with an absence of atmospheric gases, cc's are stable and survive. These cc clasts, by being incorporated into regolith have survived on earth by being encapsulated into another type of meteorite, one which has survived in the dry environment of the Sahara Desert. (Value range E).

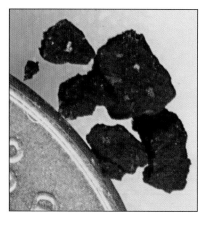

Tagish Lake, British Columbia, Canada: This carbonaceous chondrite landed on the frozen surface of Tagish Lake. The cold temperatures and careful collection techniques enabled volatile organic components present in the meteorite to be identified. These compounds included aromatic hydrocarbons, ketones, and aldehydes. Also reported are organic acids like acetic and formic acids, both found in greater quantities in Tagish Lake than in any other carbonaceous chondrite. Such compounds had to have formed in a cold environment and to have been retained at relatively low temperatures—the temperatures of deep space and an absence of any heat generating, radioactive material in its environment after it was formed in the pre-solar nebula. Tagish Lake is similar to Orgueil, a fall in France in 1856 and the first carbonaceous chondrite to be analyzed and found to contain organic compounds. Tagish Lake, like many other carbonaceous chondrites, also shows that the parent body that produced it contained water.

Unindexed NWA: The probability of a carbonaceous chondrite occurring in a meteoritic breccia is not as low as might be expected. Carbonaceous chondritic material is believed to constitute material of some dark appearing asteroids as well as comprising the dark material of comets. Fragments of either of these objects might become incorporated into the regolith of an asteroid, the source of many brecciated stony meteorites

Warrenton, Missouri: Fell Jan 3, 1877. Type III (CO3) carbonaceous chondrites are very friable and crumbly----this one is particularly so. The small specimen crumbled when placed on the surface of a flat bed scanner. (Value range G).

Organized element: Utilizing more sophisticated methods of analysis in the 1950s, the organic components of carbonaceous chondrites were more clearly identified—a situation which renewed the question as to their origin. Examination of them under high magnification using both oil immersion light microscopy and the electron microscope revealed small, regularly shaped structures resembling fossils. Aware of the implications if they really were fossils, they were given the noncommittal name of "organized elements." Later work determined that they were small crystals—crystals formed somewhat in the same manner as snowflakes (which are also crystals) form in the atmosphere. Here the crystals formed in the vacuum of interstellar space, possibly from a plasma given off by a supernova.

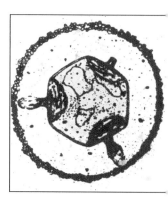

An even more significant goal was that of finding actual fossil remains of life in cc's. Both Orgueil and Murray were examined under the electron microscope for such, and minute fossil-like, symmetrical objects were found that were given the non-committal name of "**organized elements**." Recognizing the scientific implications involved if these so-called organized elements actually were of biogenic origin, most scientists remained noncommittal regarding their origin.

The second wave of interest in "extraterrestrial fossils" in meteorites more recently was associated with the discovery of "fossil nannobacteria" in an SNC, a Martian meteorite. The meteorite this time was Allen Hills, a SNC meteorite found in Antarctica in 1993. Allen Hills is an undoubted SNC, a piece of Mars thrown into space from high velocity impact on that planet's surface. This particular one fell on the ice in Antarctica and was kept in cold storage for millions of years. (The dry deserts of Antarctica preserve meteorites in a manner unequaled to any other known portion of Earth.) Examination of calcareous material filling a small crack in it revealed, with examination under the electron microscope, bacteria-like objects smaller that normal bacteria—so called nannobacteria. This find generated additional interest in examination other cc's like Murchison (fell September 1969) and Allende. Both of these meteorites, under the electron microscope, shows the presence of regular microfossil-like bodies; however, as is the case with the organized elements and Martian nannobacteria, both are best explained as being of inorganic origin. The presence of any actual fossils in a meteorite would be an extraordinary scientific event and extraordinary events in science require extraordinary verification. Occam's Razor should also come into play, where such very small, regular bodies are best explained as being crystals—crystals possibly formed in the vacuum of outer space in a manner similar to the formation of snow flakes (which are crystals) and not relics of primitive organisms from some distant time and place in the cosmos.

Glossary, Chapter 5

Calcareous Matter: Material containing calcium carbonate. In reference here to a filled crack formed in a meteorite from Mars which, after being formed lay on the surface of that planet. During this time the crack was filled in with calcium carbonate formed as a consequence of Martian weathering, probably with water conveying the calcium ions into the crack. This filled crack was the source of the minute bacteria-like objects ("fossil nannobacteria") reported from the Allen Hills, Antarctica, meteorite.

Carbonaceous Material: This material, found in terrestrial sedimentary rocks, consists of coal-like material as well as tar-like compounds (petroleum) of undoubted biogenic origin. On the earth, carbonaceous material is associated with sedimentary rocks, which contain petroleum as well as some coal-like material, all of which is of undoubted biogenic origin. The carbonaceous material in carbonaceous chondrites, in contrast, appears to have been formed from non-biogenic processes taking place at low temperatures, probably within a nebula.

Disclaimer: In natural phenomena, there is usually an exception to most statements—thus the necessity of stating these exceptions, especially when some natural phenomena involves a legal issue.

Forms of Iron: Iron occurs in three chemical states, including metallic iron, which is a form of iron having the same number of (negatively charged) electrons associated with each atom as there are positively charged protons. The other two forms of iron possess fewer elections, ferrous iron having lost two of them and ferric iron having lost three of them. To compensate for these lost electrons, the iron atoms bond with some other element and share its electrons to form a chemical compound. Ferrous iron is found in fresh meteorites; ferric iron is absent in meteorites but is present in those that have had a long residency on the earth and have reacted with the earth's atmosphere (*but see disclaimer*).

Lithification (lithified, adj.): The process by which a loosely compacted material can be cemented or compressed together. Loose, fragmentary material in space can be lithified by shock waves produced from impact or from partial melting (softening) produced as a consequence of the heat of radioactive-element decay.

Occam's Razor: An axiom of science named after William of Occam that states that the most obvious explanation for some phenomena is usually the correct one. In its usage here, an inorganic origin for carbonaceous matter in cc's is a more logical source for organic compounds than is the inference that they were produced by life existing at some distant time and place in the Milky Way galaxy.

Organic Compound: In nineteenth century chemistry, an organic compound was defined as one which was produced by living things, that is produced from the biosphere. Complications regarding this definition arose when organic compounds like those found in carbonaceous chondrites are considered. Organic compounds, like acetylene, can be formed by reacting calcium carbide with water, there being no living things involved. The twentieth century and current definition of an organic compound is that they are covalently bonded compounds of carbon and hydrogen.

Rock Flour: Finely ground material (powdered rock), which can be produced from high velocity impact.

Chapter Six
Northwest Africa (NWA) Meteorites

NWA Meteorites
—Third Class "Citizens"?

This plethora of meteorites has often been snubbed and ignored by some collectors and otherwise serious meteorite enthusiasts. This appears, in part, both because of their abundance and their low price—low prices for what are really large meteorites, but ones that, which in the rough, look much the same. In a way, they might be considered a gift of nature, or perhaps as a "democratic act" of nature—a "gift" in that nature has provided the conditions to preserve them on what is otherwise a hostile planet for meteorites, and "democratic" in that they are available (at the time of writing) at a low cost to almost anyone willing to investigate and appreciate them. NWA's are extra nice in that **they are accessible** and also because **they are real meteorites**. Some rumors having circulated to the contrary that NWA's are really earth rocks masquerading as space rocks. If still uninspired by them or by other meteorites, read the following!

The rocks and minerals shown here are different! With rocks and minerals, one generally thinks of phenomena related to **geology**—phenomena such as cooling lava or magma, sediments and sedimentary rock, fossils and rock strata come to mind. ***With the rocks shown in this book (including NWA meteorites), one has to relate to nebula, stars, and stellar explosions. With these come high levels of radioactivity, high levels of heat associated with that radioactivity, high velocity impact, and the vast, black void of interstellar space. Meteorites relate better to astronomy than to geology—they are a product of deep space.*** Meteorites are rocks, but they occupy an entirely different category of rocks or geocollectibles. It might also be added that these are the only things that really come from the "other worlds" of deep space and actually are collectible. NWA meteorites come in a number of different types, so they can become a vehicle by which new collectors can be pulled into the collecting and love of meteorites. If this happens, what often happens next is an attempt to acquire as many different types of falls and finds as possible. However, the real fascination with them is the process of "getting to know their (sometimes tight-lipped) secrets—secrets which yield clues as to the conditions under which the myriad of compositions and textures seen in meteorites were produced.

NWA Meteorites, an Extraterrestrial Bonanza

This is a "window" opened to allow the "average person" who might get interested in meteorites to now have some if he (or she) so desires.

A large quantity of delightful stony meteorites became available in the late 1990s. This was the introduction of meteorites found over a large portion of the Sahara Desert of northwestern Africa. The availability of such a quantity of "geologic" material from "other worlds" is a consequence of a series of serendipitous events. One of these is a greater awareness of extraterrestrial materials as a consequence of the space programs of various countries including that of the US, the European space agency, Russia, and China. This was coupled with a flood of quality fossils from the North African country of Morocco and the generation of Moroccan fossil dealers who were (and are) eager to incorporate a related commodity into their business in the form of meteorites. As a consequence of these events, meteorites became available in unexceeded quantities from the vast desert region of northwest Africa. A steady flow of meteorites for over a decade has supplied both science and the collector with a bonanza of material—most being stony meteorites. (The near absence of metallic NWA meteorites had been attributed to dealers removing these to supply a separate market.) Stony meteorites, although initially not as spectacular as the siderites, are the meteorites that are the most variable and also give the greatest amount of scientific information, information which otherwise could have only been obtained by a spacecraft visit to the asteroid belt. For various reasons, the collecting community has had mixed reactions to what is undoubtedly an unprecedented bonanza in the world of meteorites.

A group of small premium NWA's with some specimens showing ablation surfaces and fresh, unweathered interiors. This small lot is of relatively fresh finds, some still showing fusion crust with little evidence of weathering. These are the more desirable of the NWA specimens.

Box of NWA meteorites acquired from Moroccan fossil dealers at the 2009 Tucson mineral, gem, and fossil show. Groups of meteorites like this have been gathered from the dry deserts of northwest Africa from the late 1990s to the time of writing and offered at the major shows at Tucson and Denver. Offerings may vary, most being rusty, terrestrialized chondrites. Few earth rocks contaminate the lots—locals who collect the meteorites are quite astute in distinguishing meteorites from earth rocks. Presumably the lots of meteorites offered have already been examined for exotic types like lunars, martians or even siderites that are then offered at premium prices. When the common specimens are cut, however, a number of interesting surprises can occur.

A box of NWA's acquired at the 2010 Tucson Show. The box weights 34 lbs. Specimens are packed with Moroccan newspapers and show terrstrialization. They come with no sorting or identification. Almost all are meteorites. Very few Earth rocks find their way into these lots.

A lot of small NWA meteorites obtained at the Denver show. Different meteorite lots will vary somewhat in both the types of chondrules present as well as the amount of weathering and terrestrialization on them. Presumably some of the lots represent specimens collected from the same fall or field—different lots coming from different regions of the desert.

A few "sour notes" regarding NWA meteorites have surfaced, one being that their sale goes to the support of Muslim extremist terrorism. This the author finds hard to believe considering the minute size of the meteorite market compared to other sources of available funds like petroleum and other mineral exports of the Arab world. The other sour note is that NWA meteorites, like all geocollectibles, should remain in the region where they occur. The author has been only too aware of many excellent minerals, fossils, and artifacts either being destroyed or reburied because of a lack of interest in them by locals and "those in charge." Collectors, by offering a modest, and reasonable amount of money for these items often assure their preservation where otherwise they would be lost to both science and education.

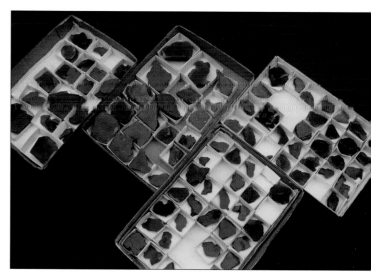

Cut and smoothed specimens are sorted and stored by lot. Part and counterpart are stored together.

Specimens are sawed using a trim saw (diamond blade) like this. The author has sawed specimens manually as that is less time consuming than using a vice feed. Loss of meteorite material can be held to a minimum when one is accustomed to sawing in this matter. Considering the large number of NWA's in a typical lot it would otherwise consume an inordinate amount of time to cut any other way. A lapidary wheel is used in finishing, utilizing a succession of progressively finer diamond laps to smooth and polish (if necessary) the NWA's.

Of concern with both collectors and science is the fact that NWA meteorites have for the most part, been undocumented as to where in the desert they were collected. Mapping positions where specific meteorites occur can enable the plotting of a what is known as a strewn field. Plotting individual strewn fields in turn, can give significant insight into the distribution, size, and frequency of specific meteorite falls during the past 100,000+ years. With desert "locals" indiscriminately collecting meteorites, such data is lost.

Less sympathetic to the author are the concerns of some collectors that this flood of meteorites will lower the value of them and thus debase the value of their entire collection. Some dealers also have the same concern with some even refusing to handle them. From an educational perspective, NWA meteorites, both as a consequence of their availability and their low cost, are allowing a greater number of persons, to whom meteorites previously were inaccessible, to now become familiar with them. Familiarity on a first hand basis with space rocks can be an interesting and rewarding endeavor and by default may also lead to familiarity with other geologic materials as well. It is the author's, (all-be-it) somewhat bias opinion, that becoming familiar, in a "hands on" manner with some natural material, like meteorites, creates a permanent and strong interest in the science relating to that material and fosters permanent learning as well. It is with this in mind that he finds it unfortunate that some persons and institutions have attempted to curtail private interest in and ownership of these (and other) meteorites.

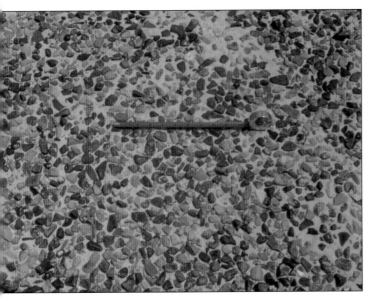

NWA meteorites can accumulate in the dry desert environment often in what is known as a lag deposit—also known as desert pavement. Fine sediment gets blown away concentrating stones which can include meteorites. Arab "locals" who traverse the desert find the meteorites and they then are purchased by Moroccan fossil dealers. A near absence of earth rocks in the lots the author has examined show both collectors (and dealers) to be highly skilled in distinguishing meteorites from earth rocks.

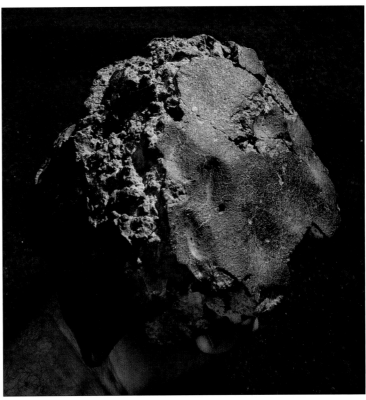

Part of the ablation surface of a (weathered) NWA "nose cone." Such sculpted meteorites (sculptured by ablation upon atmospheric entry) are both rare and desirable. They have become more accessible with the "flood" of NWA meteorites available, but specimens like this still represent only a small fraction of those collected.

Another example of a desert floor lag deposit—note that most of the stones are pretty much the same size. Sweeping over these stones with a metal detector would be a good strategy for finding meteorites. The black stones could be meteorites, a check with a magnet would verify that they were.

View of another partial NWA "nose cone."

Northwest Africa (NWA) as photographed by Apollo-17 (NASA photo). This large desert region shows well from space as it is rarely cloud covered. At the top of the picture is the Red Sea rift zone, to the left of that is the vast desert of northern Africa, the source of NWA meteorites. This area rarely has precipitation so that exceptional dry conditions preserve meteorites which fall. During the Pleistocene ice age, tens of thousands of years ago, the climate of this region was wetter than it is today. What weathering and terrestralization NWA meteorites underwent probably took place at this time.

Weathered ablation surface of a large NWA stone showing regmaglypts—depressions formed from ablation on entry into the earth's atmosphere.

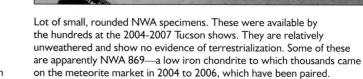

Lot of small, rounded NWA specimens. These were available by the hundreds at the 2004-2007 Tucson shows. They are relatively unweathered and show no evidence of terrestrialization. Some of these are apparently NWA 869—a low iron chondrite to which thousands came on the meteorite market in 2004 to 2006, which have been paired.

Angular specimen with edges beveled from either ablation or aeolian (wind) erosion while in residence on the earth, probably the former.

Surface of weathered NWA specimen: the fusion crust is gone as a consequence of terrestrial weathering, but the stones shape is still that developed from friction generated upon entry into the atmosphere.

Group of more angular (and weathered) NWA specimens which were available in greater quantity than those shown in previous photos.

Two NWA's with weathered fusion crust.

Various lots of NWA specimens—some lots are sorted out by degree of weathering and/or terrestrialization.

Groups of partially terrestrialized NWA's being stabilized by the application of a thin form of a cyanoacrylate adhesive (Superglue). Some NWA's are quite crumbly (friable) from partial terrestrialization. They can be prevented from coming apart before cutting by impregnation with cyanoacrylate, a resin which works especially well.

A stabilized (with cyanoacrylate) NWA specimen with weathered surface. The meteorite has been cut to show its interior (a breccia), which shows fragments (clasts) formed millions (or more likely billions) of years ago in space by asteroid or meteorid impact. The meteorite shows both chondrules and a fragmental texture (breccia). L4 chondrite (Value range F).

Upon initial examination of a group of NWA's, especially the somewhat weathered ones, most don't look too exciting, especially if they are not cut and polished. Initially what is seen looks like a group of drab, rusty looking rocks which look more or less all the same. One might legitimately question whether such ordinary looking rocks really are from "another world" and even that they are meteorites. Three major categories are represented by these rocks and these initially appear to have been sorted out either by Moroccan dealers or by secondary and tertiary dealers. The first category are the obviously unusual ones (premiums) which include metallic meteorites, the lunars, the martians (SNC's), and some odd looking stones which may or may not be of extraterrestrial origin. (Terrestrial rocks are rare in NWA meteorite lots as sorters really are good at spotting and removing them or the locals who actually collect the meteorites are good at making this distinction.) The second separated-out-group are those specimens which look like what a stony meteorite is supposed to look like—dark greenish or black rocks with specks of metal usually showing some obvious fusion crust and remaglyphs. This group is set aside for sale at the major "rock" shows such as those in Munich, Tucson, and Denver. The third group are the weathered stones—these being rusty and rather ugly looking rocks. They usually are weathered to a brownish color from terrestrial oxidization (terrestrialization) and show no obvious ablation surfaces. They look earthly. These are the ones sold at the lowest prices and the ones which the author saw at the 2006 Tucson show being more or less ignored by collectors. These rusty, oxidized meteorites as well as most of the more spectacular ones rarely have any locality information associated with them. Presumably this is because the locations from which they were collected were not documented. Considering the relatively consistent desert conditions under which they have resided on the earth (hot, dry), the amount of weathering which they do exhibit probably relates to their earthly residence time—the longer on the earth, the greater the amount of weathering.

More On the Brown, Oxidized (or Terrestrialized) NWA's

These, the most highly weathered of NWA meteorites are the NWA meteorite "underdogs." They show minimal amounts of metallic iron even when polished surfaces are examined (a process which tends to highlight the metal) and their make-up, when freshly cut, appears homogeneous. When one looks more closely, however, or when a smoothed, sawed surface is allowed to "age," diagnostic features appear that reveal their distinctive extraterrestrial origin. Chondrules, vague from perhaps hundreds of thousands of years (or more) of earthly weathering, as well as "blebs" of metal not obvious in an uncut or freshly cut specimen, now start to show up. Metallic specks present generally are small and those that occur appear to be a remnant of what has not been altered (oxidized) from terrestrial weathering. Veins of limonite can sometimes be seen where this hydrous iron oxide (a product of earthly weathering) migrated into cracks which probably existed in the original unweathered meteorite—this mineral filling in cracks formed either from extraterrestrial asteroidal impact or from impact with the earth upon landing belies their extraterrestrial origin. Oxidization products of nickel are not usually obvious, although a few specimens seen by the author do exhibit some secondary greenish, hydrated nickel minerals similar to those formed from the oxidization of millerite (See Chapter Nine).

More on NWA's, and Yet Another View!

The relatively unoxidized NWA's show unaltered, original compositions for the most part. They may also show phenomena like fusion crust, remaglyphs, and cone-shaped ablation surfaces formed upon entry into the earth's atmosphere. They are nice! The majority of these NWA meteorites, like most meteorites, belong to that large group referred to as stony meteorites. This meteoritic window of opportunity opened around 1998 and has seen a steady flow of meteorites with their sale and distribution encouraged by Moroccan fossil dealers. Around 1998, Moroccan fossil dealers, already familiar with the "rock" business, began to interact with locals of southern Morocco, Mauritania,

Sliced specimen showing junction between two well joined and distinctive clasts. Many stony meteorites are breccias, like this specimen. It possibly formed on the surface of an asteroid as a regolith. Impact metamorphism at a later time can weld the fragments together.

A small specimen showing distinct clasts, the bottom clast is composed of light colored silicates, upper one at 2:00 is almost black and is possibly a **carbonaceous chondrite**. The small circular structures are chondrules.

Same specimen as in previous photo using different lighting, which shows metal component. Note parallel lines across the sawed surface. This is a consequence of grinding the cut surface after cutting to remove saw marks. The face has not been polished. If polishing were done, the polishing lines would not be present.

L4 with a granular texture. The brown areas are the location of small masses of weathered nickel-iron surrounded by a halo of hydrous iron oxide formed after the fall of the meteorite. Ghost-like clasts are evident. (Value range F).

Libya, and Algeria, and encouraged them to collect the dark rocks found sporadically and in clusters over this large area. This evolved into the window of opportunity currently available to acquire meteorites (sometimes large ones) at prices previously only dreamed about. Thousands of these NWA meteorites have been dispersed through the large rock, mineral, and fossil shows like those in Tucson, Arizona, and Munich, Germany. As this is written, NWA meteorites remain at a low price range as a consequence of the following:

The clasts of this breccia show distinctly in this sliced specimen (part and counterpart). Note the clast at the right side of the slice on the left. It lacks chondrules and has a fine grained texture. To the left of this clast is another darker clast possibly from a carbonaceous meteorite. Metal does not show at the angle the photo was taken. (Value range F).

The same sliced specimen shown in the previous photo (but now positioned to bring out metal). Note again the texture-less clast to the right of the slice on the left. The angle at which this photo was taken **shows metal**. Note also the metallic globule butting against the clast. The clast was in existence when the metal globule was still molten, so this clast must be particularly ancient and was **not** part of a later regolith as are many meteorite breccias. (Value range F).

1. A saturation of NWA's in the meteorite market with so many specimens coming upon the market so quickly (at least for awhile at the shows).

2. A view by many meteorite collectors that NWA meteorites are "substandard" and thus are less desirable as they lack providence in the form of specific locality information. Lack of a pedigree and cachet of respectability which adorn the classic meteorites detracts from them. They are believed by some collectors also not to be meteorites at all but rather are earth rocks which resemble meteorites.

With regard to a lack of specific locality information, it might be mentioned that a major concern for a meteorite's residence time and location on the earth is somewhat inconsequential and in some ways even somewhat silly. It is like having to know exactly in what particular room a bottle of wine was stored and where it was purchased rather than from what country the wine actually originated. What part of the Solar System the meteorite is originally from is of greatest interest and importance. Unfortunately with most meteorites such information is currently unclear. What a meteorite has to "say" regarding its place of origin and the conditions which existed when it formed—deep in the abyss of geologic (or astronomical) time—is also important (but unfortunately, like the former can also be difficult to determine). The appeal of a meteorite is really about what it represents, meteorites being the only tangible object from space that one can acquire and collect. A caveat for desirability of these meteorites may also reside with the large numbers collected. All known meteorites (currently) are believed to have originated from the Solar System (although some components in them were produced before the Solar System formed). The possibility exists that a small portion of meteorites might exist that are of extra solar origin, originating from some other part of the Milky Way galaxy. An extra solar space rock would require a long period of time to make an interstellar journey; however, with geologic materials, time is of little consequence. The type of meteorites that might be extra solar would probably be chondritic types, possibly almost indistinguishable from chondritic examples formed from the solar nebula. Probably all nebula capable of producing type G stars like the sun are capable of producing similar chondritic type meteorites. The large number of specimens of NWA meteorites collected would increase the possibility that a small portion of this material might just be of extra solar origin. Who knows? Currently, no one!

Pairing of NWA Meteorites

Pairing is the assemblage of two or more meteorites from a single fall. Under ideal conditions the location of a meteorite is plotted on a map and if enough individual specimens of a single, fragmented meteor are located and plotted, the plotted points define an ellipsoid known as a strewn field ellipsoid. In this configuration one end of the ellipsoid contains the larger specimens and the other end the smaller ones. The reason for this pattern is that the larger specimens traveled to the point of impact faster than did the smaller ones—larger fragments having been reduced less in their cosmic velocity in their encounter with atmospheric friction. In other words, the larger fragments have a more direct, straight line trajectory than do the smaller fragments which, being more affected by friction, are slowed down more, with their trajectory now being influenced more by the earth's gravity. The smaller fragments, deviating more from a straight-line-trajectory that the larger ones, result in the geometric pattern of an ellipsoid—smaller specimens at one end of the ellipsoid and larger ones at the other end.

Some purists bemoan the fact that locals, collecting meteorites, fail to plot where they find them; as a consequence of this, some significant scientific information is indeed lost. If there is a valid argument against the local collecting of NWA's, it is this loss of scientific data; however, the author is of the opinion that such data loss is compensated for by the availability of these meteorites to a much larger group of persons. The availability of NWA meteorites enables persons who otherwise might not even see a meteorite to now own one with only a nominal outlay of money. The end result of this availability results in a greater overall interest in meteorites by a greater number of persons.

The same slice as shown on the right of the previous photo—the specimen has been angled to reflect (and show) the metallic component. Notice that a distinct metallic globule penetrated the clast; this is the same globule which butts against the clast shown in the previous photo. Note also the spherical metallic globule below and to the left of the larger globule.

An H3 chondrite in which chondrules are barely visible. The grey lineation is where the meteorite has been bonded together. A clast to the left of the repaired crack has been outlined in blue. (Value range F).

The same slice as in the previous photo (but now turned at 180 degrees) with metal subdued. Note the brownish layer at the (now top) of the clast just below the I in Iron (ferrous) phosphide. This is a layer of schribersite, a iron-nickel phosphide, a mineral exclusive to meteorites.

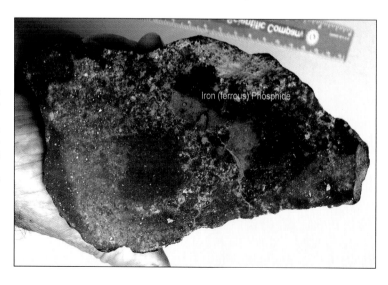

An L4 with chondrules barely visible. Note grey clast at top (more on this later). This face has not been polished.

An H3 chondrite with some conspicuous chondrules. Note the almost perfectly spherical chondrule just to the left of the C in chondrule. Metal in this image shows a slight pinkish tinge.

Four sliced, oxidized chondrites like those seen above. This specimen was photographed just after being cut. Chondrules and clasts (as well as other textural features) show up better on cut faces after the face has been exposed to the air for a few months).

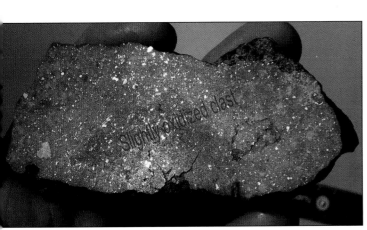

H3 chondrite. Note the "slightly oxidized clast" to the right of the label. Hydrous ferric oxide (rust) stains this clast. Note clast at bottom left also (to the left of the word slightly). (Value range F).

An oxidized chondrite under normal contrast. This specimen was allowed to "age" for a month after cutting and before being photographed. (Value range F).

Somewhat oxidized chondrite with numerous small chondrules. This is an oxidized chondritic meteorite—part and counterpart. These specimens, showing distinct chondrules, are distinctive among oxidized NWA's.

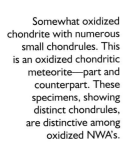

The same specimen, but with higher contrast. Note that the chondrules show up better.

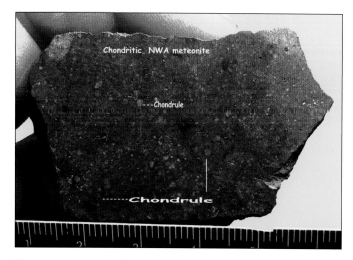

The same specimen as above with less light and some of the chondrules labeled.

Similar H4 with pronounced armored chondrule. Chondrules often are coated with a metallic sheath or armor (armor-plated chondrules). This was done by some sort of natural "sputtering" process. The coating of an object with metal deposited in a vacuum is known as sputtering. This phenomena was utilized in the manufacturing process of duplicate sound recordings. In this process the master recording is placed in a bell jar-like container where electrodes made up of some metal give off (sputter) metal which armor coats the master record. This is later backed up with more metal to be used in making copies of the master. The metallic armor coating of chondrules was a similar, natural process where the source of the metal (instead of electrodes) being a red giant star or a nova or supernova—the chondrules being, instead of a master record, the objects being coated with metal. (Value range F).

A black, unweathered, fresh, and almost texture-less, metal free specimen under higher contrast. Under these conditions granulation (rock flour?) and chondrules are visible.

L6 chondrite with some cavities (vesicules?) at upper right. Chondrules can barely be seen in this slab which as been subjected to thermal metamorphism. This has all but destroyed the chondrules, a phenomena found in many chondritic meteorites.

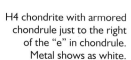

H4 chondrite with armored chondrule just to the right of the "e" in chondrule. Metal shows as white.

The same oxidized chrondritic specimen as in a previous photo showing a somewhat vague, oxidized clast labeled just to the right of the "T" in clast.

Large slab (28 cm X 32 cm) cut from a 15 Kg NWA meteorite. This is a metamorphosed chondrite in which the chondrules show up as armored chondrule "ghosts." The parent mass probably reached a temperature of at least 800 degrees C, not hot enough to melt it but thermal metamorphism has all but wiped out the chondrules. A group of armored chondrule "ghosts" can be seen to the left and below the "G" in Group. Larger ones can be seen below the penny. The outline of the chondrules has been preserved as the chondrules, after being formed, were then given a thin coating of metallic iron sprayed (or sputtered) perhaps by a dying, red giant star or a nova. Look carefully at these slabs and you will see many other chondrule "ghosts."

Small specimen with (round or spherical) chondrule and rectangular (carbonaceous?) clast. Note larger clasts to the right.

Same specimen as in previous photo but with less contrast. Features in NWA meteorites (as well as in others) show up better with more contrast.

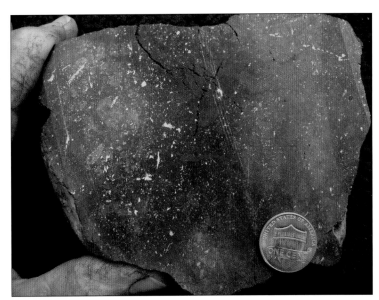

Cut surface of NWA acquired at a rock show. The specimen is similar to the above specimen; probably paired with it. This specimen however shows more metal. The specimen was cut on a 24", automatic feed saw. One surface was polished, however the texture (and metal) shows best on the unpolished surface. This photo shows this unpolished surface just after polishing the counterpart (note where the metal has spread over the meteorites surface from the saw blade, smoothing and polishing will remove this). Note also the dirty, iron-stained fingers as a consequence of cutting and polishing meteorites. Working with them can be messy and they leave a lot of iron stains—almost like working on old cars but without the oil. (Value range D).

Note cluster of ghost chondrules in center of photo. A large number of armor-plated chondrule ghosts are evident here—look close! Slab cut from the same 15 KG meteorite as slabs shown above.

Another large slice from this L6 chondrite—chondrule "ghosts" can also be seen but are not obvious, but if you look close a lot of them can be seen.

Slice from the same meteorite as in previous photos (but cut from opposite end). Some "ghost" chondrules can be seen by their preserved coating of metal (armored chondrules)—also some larger chondrules of a different type (grey chondrules) are just barely visible but were not armor-plated.

Large cut slab with numerous small metallic specks, and chondrule ghosts. Note cracks from terrestrial weathering in a hot, desert environment.

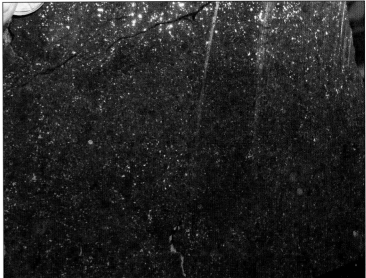

More chondrule ghosts. The vertical, slightly curved lines are saw marks (spreading of meteoritic metal) from cutting the meteorite on a 24 inch rock saw. Polishing such a surface appears to reduce some of the features like ghosts but brings out other features like clasts.

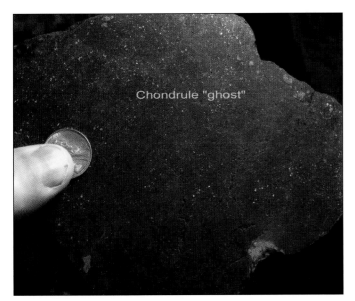

Chondrule "ghost"

Note chondrule "ghosts" directly below the word ghost. Another slab cut from 15 Kg NWA meteorite.

Another slice from this large NWA specimen. Chondrule ghosts are many, but hard to spot at first.

NWA Meteorites-II

Of the large number of meteorites which have come form the Sahara Desert, the majority have little or no locality information available; in some ways of thinking these are considered "orphaned meteorites." Such a "second or third class" relationship of NWA's to the classical meteorites, which have well established earthly provinces and pedigrees, has resulted in their low acquisition cost as a meteorite or geological collectible. Because so many NWA's have come onto the meteorite market, and because stony meteorite classifications often don't appear to apply to them very well, the proposed classification scheme is presented. This can be especially cogent as the majority of NWA meteorites, unlike the "classics," are without locality information and yet they have distinctive characteristics. This scheme also takes into account the fact that NWA's generally exhibit some weathering; however, compared to meteorites found in wetter regions of the earth (with much shorter earthly residence times), this weathering is minimal. In this regard, it is the author's opinion that weathering is consistent in its effects over this large region—a region which is consistently quite dry, the degree of weathering being primarily influenced by a meteorites residence time on the earth's surface. In particular, the below scheme references the iron and magnesium silicates (ferromagnesian silicates) as these appear to be the components most responsive to weathering in this dry environment. These silicates are converted, in part, to hydrous iron oxide (either amorphous as limonite or to the somewhat crystalline goethite) and the amount of conversion is assumed (all other things being equal) to be a function of the residence time on the desert surface, the more ferric iron oxide in the meteorite, the greater the residence time. This scheme is proposed to be used as a first order classification using visual examination. Microscopic examination and thin section work would produce more precise information as well as a more conventional determination. In this scheme **L stands for low metallic iron, H stands for high metallic iron** as with other chondritic meteorites. The condition of the meteorite's interior refers to the presence or absence of hydrous iron oxide. Weathered=present, unweathered=absent.

Starry night—in reflected light—with high contrast. Metal shows as white. This is an H chondrite. (Value range C).

A Starry night—as above but in reflected light.

Starry night, as above but with less contrast.

L4 with oxidized metal (ferric oxide halos).

L4 with hydrous ferric oxide halos around numerous metal inclusions. This type of alteration can happen even with falls in a matter of a few decades so that residence time for this NWA meteorite was probably short.

Part and counterpart of weathered H5. (Value range F).

Polished H6 chondrite, a very fresh specimen with no oxidization. Chondrules and clasts are vague from thermal metamorphism—the meteorite somewhat resembles basalt—but a basalt with metal specks. (Value range F).

Single view of the previous image. Chondrules are not evident, but this is still a chondritic meteorite, as are most NWA's, the chondrules having been totally destroyed by thermal metamorphism.

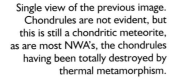

Fresh and unoxidized L5 in reflected light to show metal.

L5 with metal oxidized around metallic inclusions. Note vertical linearity, a phenomena usually not associated with chondritic meteorites. (Value range E).

Sliced, part and counterpart of same specimen as in previous photo, metal more apparent in this view.

L5, the light has been positioned to minimize metal. Chondrules are hard to detect, but they are there!

Unoxidized (or terrestrialized) L5 with metal showing. (Value range F).

Group of terrestrialized L4's photographed shortly after being cut.

Terrestrialized L4 with conspicuous limonite vein. Limonite is hydrous ferric oxide, a mineral formed from meteorite weathering on the earth. The crack has been filled with limonite, which may have formed in space from impact with another extraterrestrial object or it could have been formed when the meteor hit the earth to become a meteorite.

Terrestrialized L4 with (ugly) lawrencite disease. A long residence time on the earth did not remove the ferrous chloride, the cause of meteorite disease—in fact it often *seems to make it worse*.

Terrestrialized L4 with limonite filled cracks, which were probably formed before earth entry.

H5 at a lighting angle which shows metal—note armored chondrule at upper left. There is another one to the right of it but it does not show as it has no metallic armor. There is a saw cut to the right but it does not show up conspicuously at this angle of light.

Terrestrialized L4 with numerous secondary limonite veins shown in reflected light.

Same H5 specimen as above in low angle light to show metal. Note the appearance of chondrules to the left, both now show up.

A large number of chondrules make up this L3. Note conspicuous large grey one at the top left. At the top right is also a dark clast.

Same specimen as above at higher light angle. Note chondrules at top (left of center). The line at the right is the saw cut.

Large chondrule(?) near center—clast with small metallic inclusions at mid-bottom. Note elongate clast at middle left.

Note two large grey chondrules(?) at upper left and a large vague clast occupying the bottom of the cut surface. (Value range E).

Very round (circular) chondrule at bottom. Right: this one is distinctive from the other numerous chondrules which make up this H4. Many different types of chondrules occur, the specific conditions which produced different types of them are puzzling. Meteoritic scientists who work with chondrules refer to them as being "tight lipped."

Proposed NWA class. (Type of chondritic meteorite) condition of sliced interior:

NWA-I L 3,4 chondrite, low in Fe highly weathered

NWA-II L 3,4 chondrite less weathered

NWA-III L 4,5 chondrite unweathered interior

NWA IV L5,6 chondrite unweathered interior

NWA-V H 3,4chondrite weathered, clasts distinctive

NWA VI H 4,5chondrite unweathered, clasts vague

NWA VII H 4,5chondrite some oxidation, chondrules present

NWA VIII H 4,5 (starry nights) unweathered, clasts vague

In this scheme NWA I's are the brown, rusty looking and highly oxidized (or terrestrialized) meteorites which initially seem uninteresting but which are some of the cheapest and most readily available. On an unsawed, rough surface, they show no features. On a sawed surface they may show vague chondrules, which for some reason stand out a bit more as the specimen "matures" in a collection.

1. NWA's II are similar to NWA I, but their interiors show more features, particularly better outlines of clasts and chondrules; they have minimal metallic iron.
2. NWA III's show clasts and chondrules but **not** so well as with freshly fallen meteorites (which usually show textural details with greater detail).
3. NWA IV's have essentially unweathered interiors, but unlike NWA II's show only vague chrondrules. Here the chondrules have been almost obliterated either by *impact* or **thermal** metamorphism and are **not** vague because of terrestrial weathering.
4. NWA V is an H 3 chrondrite, a stony meteorite with more metal than the L 3's. The interior of this class may be somewhat weathered which makes the remaining metal less obvious (some of the metal {nickel-iron} of this class has been converted to hydrous iron oxide). **Clasts will be fairly obvious** in this meteorite as are the chondrules.
5. NWA VI This is an H4 or H5 chondrite with little evidence of **chondrules** as they were modified before entrance onto the earth by impact or from thermal metamorphism. **Clasts** will be discernible as in NWA V; they are not affected by weathering or terrestrialization.
6. NWA VII is an H4 or H5 chondrite in which the **chondrules** are not obvious, either from metamorphism or from weathering ... or both!
7. NWA VIII. This is an H4, H5 chrondrite in which the metal is abundantly disseminated throughout, forming a myriad of small specks. This disseminated metal in a sliced specimen resembles a star-studded night sky. These have been informally referred to as a "starry nights" after the Vincent Van Gogh painting of that name.

Numerous somewhat equal sized chondrules compose most of this L4 meteorite. (Value range E).

Same type (L4) as shown previous but with smaller chondrules.

Place Names and NWA's

NWA meteorites as presented in the previous classification scheme, are distinguished not only by texture (petrologic classification) but also by their general appearance with regard to their state of weathering (which reflects their residency time on the earth in the dry desert). This is in part a consequence of the fact that the usual designation as to where a meteorite was found or was seen to fall is unavailable for NWA's. This is because of the following:

1. Few identifiable cultural landmarks are present in many parts of the dry Sahara.

2. Collectors have not (or will not) document the place or places where specific meteorites have been found or are found.

The standard megascopic classification system is utilized for NWA specimens illustrated here. The utilization of magnification and particularly the use of thin sections and polarized light (crossed Nichols) would allow for a considerable amount of additional information and would also "nail down" the specimens more precisely; however what is presented here (and in other parts of this book for that matter is a first order, visual look at meteorites). The more specific examination of them can come later.

Group of (slightly) terrestrialized specimens that came in the same lot. Are these all from the same strewn field? (Value range E for group).

Four samples from the above lot shown in reflected light so that metal shows conspicuously.

Bottom specimen (part and counterpart) of previous photo, somewhat terrestrialized. (Value range F for both).

Sliced larger samples
from the same lot as in
previous photo.

Weathering vs. Impact and Thermal Metamorphism

Meteorites have distinctive and definitive textures which record phenomena (either astronomical or geological) not observed in earth rocks. This includes a number of textural phenomena, but what stands out specifically are the chondrules. Chondrules are a phenomena **unique** to stony meteorites and link them to their origin in a nebula; therefore **chondrules are an astronomical phenomena. NWA meteorites are predominantly stony meteorites and most stony meteorites contain, or did contain, chondrules. Some stony meteorites, upon being sliced and examined, lack chondrules and never had them.**

These are the achondrites; they are derived from planet-like "worlds."

Terrestrial weathering processes diminish textural features, which includes the chondrules. The oxidation of ferrous silicates, which takes place during a meteorites residence on the earth, can soften textural features, including both chondrules and clasts. Chondrules also can be vague as a consequence of impact metamorphism. When large masses of matter collide, like two asteroids, a shock wave travels through the colliding fragments, rearranging (or partially rearranging) the space lattice of crystals as well as imparting a kind of shock induced "molecular movement," which affects texture. This is a phenomena known as impact metamorphism and its effects either reduce textural variations or remove them altogether.

Sliced surface and weathered face from the same lot as those above, but with little metal present. (L4). (Value range F for both).

Group of terrestrialized H4's from the same lot as above.

L4 with numerous vague chondrules. Photo taken after specimen was cut. It has not been polished. This specimen and the following three others came from a lot acquired at the Denver show in 2009.

An unweathered H5 chondrite where most of the chondrules are gone or are vague—they were subdued by thermal metamorphism. An especially large number of chondritic meteorites have come from NWA. Very few of these have been dated by radiometric age dating. Those that have give whole rock dates of around 5.0 billion years, the time of consolidation of the Solar Nebula but prior to the formation of the planets. There is a possibility that some chondritic meteorites on the earth may have originated from somewhere else in the Milky Way galaxy. All chondritic meteorites dated so far give whole rock dates of 5.0 billion years. Radiometric age dating which gave whole rock ages either considerably older or younger than 5.0 billion years would suggest that these particular meteorites came from elsewhere in the galaxy. Few NWA chondritic meteorites out of the vast number which have been collected have been age dated so that there is an opportunity to find a spurious age date in perhaps a few of these. Fresh specimens like shown here would be best for doing this. (Value range E).

The other phenomena which removes (or partially removes) chondrules is thermal in origin. After condensing from a nebula, parent meteoritic material is high in radioactive material. Radioactivity produces heat and this heat can "cook" and remove original textures—including chondrules, especially when material with radioactive isotopes "clumps" together. The larger masses of material accreted from a nebula can totally melt and produce a planetoid, the smaller masses (planetismals) don't completely melt but are affected by the near melting point temperatures reached. This thermal metamorphism effectively removes or partially erases original textural features like chondrules.

Brecciation and Asteroid Impact

Most stony meteorites exhibit what is known as a brecciated texture. A breccia is the phenomena of "a rock within a rock." A breccia will be made up of angular fragments, the fragments being produced by some, usually energetic, process like volcanism (volcanic breccias), earthquakes (fault breccias) or impact breccias—the source of the brecciated textures in meteorites. Specifically in meteorites brecciation (or a brecciated texture), originates from high velocity impact as a consequence of one asteroid hitting another. Shock waves, produced by this impact, can result in the following:

1. *Loose surface material being "pressed together" with the result being a polymict breccia derived from an asteroidal regolith.*
2. *Crystalline structure or the boundaries between crystals being softened or even removed by shock metamorphism.*
3. *Original structures such as chondrules being partially or completely erased.* (This can also happen by thermal processes but the effects are different.)
4. *Granular material (rock flour) being produced in quantity.*
5. *Clasts or irregularly shaped rock fragments being produced, which may be intermixed with rock flour.*

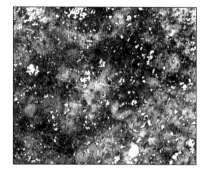

Close-up of the texture of the previous shown specimen.

Another fresh NWA H5 chondrite—a "starry night" (Value range C).

An L4 with numerous circular areas—once the site of large chondrules? Note the "twin" at bottom left just above the finger nail. To the right of that is a vague, altered clast composed of light grey silicates.

An L4 chondrite with distinct and well developed chondrules. This specimen of a particularly nice and fresh chondrite came with a lot of "premium" NWA's. (Value range E).

6. *Impact at a later time may lithify or "press together" this loose material going back to number 1—and the sequence can repeat itself.*

7. *A meteorite texture may record the previous phenomena multiple times in the form of what is known as a polymict breccia.*

A NWA
Nomenclatural Suggestion?

Meteorites are usually referenced by a place name (usually a town) near where a particular fall occurred or where the specimen originally was found. This simple and effective nomenclatural procedure works well in populated areas, but in unpopulated or sparsely populated locations pronounced or named natural features can form the basis for a name. With NWA's, a lack of documentation by the original finders as well as an absence of reference points over large parts of the Sahara Desert, which might be used in their documentation, are both absent. Currently some (really only a few) NWA meteorites have been indexed in the Meteoritic Societies' "Catalog of Meteorites." Considering that a large quantity of these desert finds have gone into either dealers or collectors hands, it might be reasonable that these meteorites, when they are referenced in the future, be referenced in terms of the original collection from which they came. With fossils, a precedent of this type already exists. With specimens from specific collections, particularly if the collection was extensive, the name of the collection from which the fossil came is used. Thus a NWA meteorite in the collection of the author might be referenced as Stinchcomb-127, etc. This is especially appropriate with meteorites, as the place where they are found on the earth's surface is not the location where they originated, which is usually the case with both fossils and terrestrial minerals. In the future, when the desert source of these meteorites dries up, reference to them using specific collections appears to be a logical nomenclatural strategy. It might also encourage collectors to donate their collections to appropriate, dedicated meteorite repositories.

What has been shown in this chapter so far represents the normal run of NWA meteorites. Occasionally specimens of NWA's surface that are distinctly different from the normal run of chondrites. Some of those are shown below.

Close-up of uncut end of previously shown specimen—note greenish cast—a very thin film of a secondary nickel mineral covers the meteorite's end.

A somewhat terrestrialized L4 with meteorite disease. The more weathered, low iron NWA specimens sometimes have this malady. Note the greenish tinge at the bottom of the uncut portion at bottom right. This is a coating of a secondary nickel mineral formed from weathering of the meteorite. Similar green nickel minerals are found on earth, which may have been derived from extraterrestrial sources; these form the subject of Chapter 8.

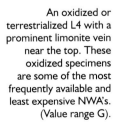

An oxidized or terrestrialized L4 with a prominent limonite vein near the top. These oxidized specimens are some of the most frequently available and least expensive NWA's. (Value range G).

Oxidized L4 in which chondrules are evident at the top but vague (Value range F).

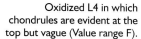

Specimen from the same lot as in previous photo with lawrencite disease.

Another L-4 with lawrencite disease, which possibly came from the same strewn field as did the previously two specimens. All came from the same lot of NWA's.

Small marble-sized sphere made from H4 NWA meteorite which is similar to those shown above. (Value range F).

Slice through H5 shows vague mottled texture. The cracks were probably formed during residence on the earth. (Value range F).

A terrestrialized meteorite from the same lot as those shown above but with numerous cracks. Some of these fractures have been filled naturally with limonite while weathering on the desert floor, possibly when northwest Africa was wetter some forty thousand years ago. (Value range F).

Another sliced terrestrialized NWA.

Both chondrules and clasts show distinctively on this L5.

Has a good weathered fusion crust—specimen has a homogeneous texture except for two large, somewhat vague chondrules and small cracks.

A fresh NWA meteorite from a different lot. Contrast this specimen with the terrestrialized specimens shown in previous photos. Note part of a lighter colored clast at the upper left. Also note small dark clast at bottom left—possibly a clast from a carbonaceous chondrite. Contrast this with the light, granular clast to its right (Value range F).

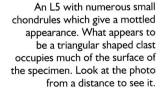

An L5 with numerous small chondrules which give a mottled appearance. What appears to be a triangular shaped clast occupies much of the surface of the specimen. Look at the photo from a distance to see it.

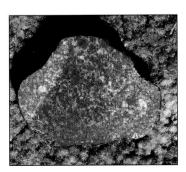

Note light colored clast at the bottom, dark clast at the top. This may be a carbonaceous chondrite clast. The meteorite is an H5, it has been metamorphosed, the carbonaceous chondrite(?) clast may have retained its distinctiveness from its containing carbonaceous matter. During metamorphism of terrestrial sedimentary rock, especially shale, with carbonaceous material present, the rock will be considerably less altered than will similar rock without carbonaceous matter. The author has collected trilobites from carbon (graphite) bearing metamorphic rock where overlying and underlying beds lacking such carbonaceous material, had the same fossils, but the fossils were almost totally obliterated by metamorphism—carbonaceous material acts as a retardant to metamorphism. (Value range F).

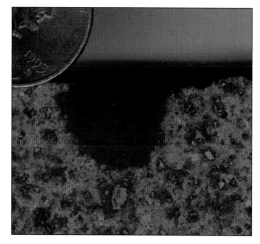

Close-up of the dark clast shown in the previous photo. Note that the clast (unlike surrounding material) is very homogeneous and fine grained.

Note fine grained, dark clast at the top.

Two distinct clasts show up here, the left one is light colored, that to the right is possibly part of a carbonaceous chondrite.

Numerous chondrules and a distinct clast on the upper left comprise part of this NWA specimen.

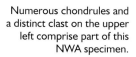

Another view of the same specimen as above.

Many chondrules and clasts composes this interesting NWA specimen. (Value range E).

Fine-grained "starry night" but with metal not obvious. (Value range E).

A large rectangular clast and smaller clasts at the top characterize this specimen. The whitish cracks are of recent origin, formed after the meteorite landed on the Earth.

Specimen shown above photographed in reflected light so that the metallic component shows.

Filled fractures characterize this H5 specimen. (Value range E).

Numerous chondrules compose this specimen.

A mottled texture characterizes this specimen. (Value range E).

Large chondrule at far left and armored chondrule at right in this H5.

An L4 almost entirely composed of small chondrules. The chondrules in this specimen show up well even on a roughly cut surface like this. Most NWA's don't show chondrules so distinctly as this even when polished. (Value range E).

Mottled texture similar to previously shown specimen. Note uneven distribution of metal and silicate areas.

NWA 2142: Brecciated eucrite. An example of an indexed NWA meteorite (indexed in the catalog of meteorites). Eucrites often are similar to basalt. Besides darker clasts, this specimen shows at 10:00 a pattern of interlocking calcium plagioclase (white) and pyroxene (dark) crystals. Eucrites and diogenites (both achondrites) can be readily distinguished from chondrites. Moroccan dealers separate them from the chondrites and sell the (obviously) rarer eucrites at a premium price. Such "premium" priced meteorites are the ones more likely to be indexed as is the case with this specimen. (Value range D).

Light colored silicates make up this H5 interior. The meteorite must be a fairly recent fall as it has a black fusion crust.

Another slice of NWA 2142 showing intergrowth of crystals of white plagioclase crystals and black pyroxene in a clast.

Brecciated eucrite, large slab. This specimen probably came from the same fall as NWA 2142; however, the larger specimen came with a lot consisting of unindexed NWA's. The fact that this is a breccia is more obvious with the large slab than with the smaller. (Value range C).

Gold Basin: Chondritic meteorites like this are similar to those from NWA, but are found in southern Arizona. Known as Gold Basin meteorites, they have been widely distributed among collectors. Could these meteorites be part of a large fall of chondrites which fell approximately one million years ago? They are similar looking to many NWA's preserved under similar dry conditions half a world apart. (Value range F).

Uncommon and Unique NWA's

Collectors of meteorites from the deserts of northwest Africa are thorough in their recognition and separation of meteorites from earth rocks. Moroccan fossil dealers purchase their finds, which include meteorite types different from the chondrites, the meteorites which constitute the bulk of what is collected. The dealers then go through the purchased lots, spotting those which, without being sawed, appear distinctive for one reason or another. These special meteorites are then sold at a premium price to collectors or

to other dealers at the major rock shows of Tucson, Denver, and Munich. It is in this manner that both lunars and martians have been found. Prices of undocumented "premiums" are considerably greater than the general run of chondrites; however, to document that a specimen really is a martian or a lunar requires sophisticated analysis including trace element distribution, atomic absorption, and use of the electron microprobe—all of which can be both time consuming and require expertise to operate and interpret. If these techniques show that a specimen has the prerequisite signatures to be a unique meteorite, the specimen is given a reference number and cataloged. Once this is done, usually half of the specimen goes to the owner dealer, the other half going to science. Such documented specimens, when placed on the meteorite market fetch considerably more than they would have before being tested, often over $1,000/gram. Some of the specimens shown here represent specimens purchased from Moroccan dealers—they have not (at the time of writing) been evaluated by any sophisticated means, however the author speculates in the captions what they might be. A few of these speculations are rather "way out."

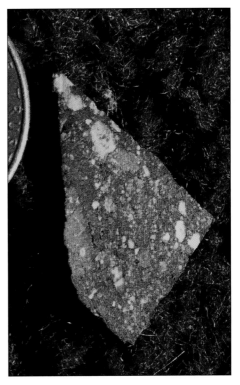

NWA-2995: Lunar meteorite. Basalt-bearing anorthositic, fragmental breccia. Fifteen lunar meteorites have been found to date. This is a slice of a NWA specimen found in Algeria in 2005. Its interior reveals many light grey and feldspathic clasts in a dark, fine grained matrix. (Value range B).

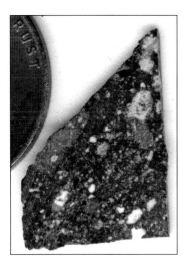

NWA-2995: Flip side of this NWA lunar slab under higher contrast than that in the previous photo. These NWA's have been well documented as coming from the moon. Their lunar origin has been determined from the following:
- Similarity in lithology (appearance) to lunar basalts brought back by Apollo explorations.
- A similarity in trace element content with Apollo lunar samples.
- Radiometric age dates (4.0 billion years) which are essentially the same as uranium-lead radiometric age dates obtained from lunar mare basalts brought back by the Apollo missions.

NWA 2995: Another view of an undoubted, documented lunar. This is one of the "jackpot" premium specimens offered by Moroccan dealers.

This specimen came with the same lot as specimens in the previous photo. Note the similarity of the chondrules. Note also the "large" mass of olivine crystals appearing as a yellowish-green mass. These were labeled as being carbonaceous chondrites (CV3's). They have large chondrules and are fairly hard. These cc's rarely are recovered as finds, however, the extremely arid conditions of the Sahara might be capable of preserving cc's from weathering. (Value range ?).

A slab from another specimen that came with the same lot of "premiums" as above specimens. Many nice chondrules here! The small slab as been moistened to bring out texture. (Value range ?).

This slab is probably from an earthly (terrestrial) rock. It came with the same lot of "premiums" as did the above specimens. Small fragments in it appear to be quartz, and quartz is rarely found in meteorites. Also the grains in it appear somewhat rounded, something else not found in meteorites (at least meteorites from *our* Solar System).

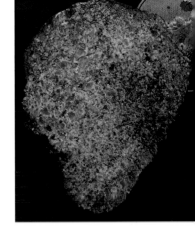

This meteorite came with a lot of "premiums." Its being a meteorite was at first questioned. However, on cutting, the distinctive signature of chondrules became apparent; it is also attracted to a magnet. It was selected by a Moroccan dealer for both being so hard (and glassy looking) as well as for its lower-than-normal amount of magnetic attraction. When it was cut, it showed numerous, well formed chondrules—though a chondrite, it still looks somewhat peculiar. (Value range ?).

This one, when first cut, was hoped to be a SNC—it suggested Nakhla (see Chapter Five). A much better match, however, is with a eucrite. It appears identical to NWA 2142; it may be part of the same fall, that is it may be paired with NWA 2142. (Value range E).

This one definitely is a breccia—probably a regolith breccia. The question is on what heavenly body did this regolith originate? Did it come from an asteroid or from a planetary body like the moon? Like many "premium" NWA's, it is not magnetic. It is a candidate for being a lunar. It is probably a brecciated eucrite but it will eventually be tested as to being such in the future. (Value range ?).

This meteorite looks similar to the previous specimen, however it is magnetic and exhibits vague chondrules. It also includes part of a weathered "nose cone" which was formed from ablation during atmospheric entry. (Value range D).

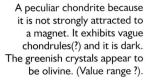

A peculiar chondrite because it is not strongly attracted to a magnet. It exhibits vague chondrules(?) and it is dark. The greenish crystals appear to be olivine. (Value range ?).

This is the source of the previously shown slab. Its surface exhibits what is either a wind blown (aeolian) eroded surface or it is an abraded surface of what originally was fusion crust. It came with a group of premium NWA's. (Value range ?).

Magnetic goethite: Rarely are hydrous iron oxides of earthly origin magnetic (brown magnetic recording tape being a man-made example). This is a chunk which is magnetic and as a consequence, it found its way into a box of NWA's.

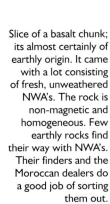

Slice of a basalt chunk; its almost certainly of earthly origin. It came with a lot consisting of fresh, unweathered NWA's. The rock is non-magnetic and homogeneous. Few earthly rocks find their way with NWA's. Their finders and the Moroccan dealers do a good job of sorting them out.

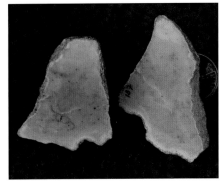

Chert. This piece of cryptocrystalline silica (quartz) is a common sedimentary rock of earthly origin. It came with a box of NWA's.

Glossary, Chapter Six

Breccia: A rock (or meteorite) made up of angular fragments (clasts). Breccias are produced by some high energy environment; with meteorites the clasts are produced from the high energies of impact.

Brecciation: The process of producing a breccia. High velocity impact is the mechanism responsible for brecciation in meteorites and impact breccias.

Clast: Angular fragments within a breccia.

Chondrule: 2-8+ mm spheres characteristic of and found only in chondritic meteorites. Chondrules were formed by poorly understood processes taking place in a nebula.

Hydrous Iron Oxide: Ferric oxide which contains some chemically combined water (essentially rust). In nature (and terrestrialized meteorites), two forms occur—limonite and goethite. Limonite has little structure, while goethite can have a fibrous structure and is generally more solid. Many NWA stony meteorites contain hydrous iron oxide—a product of reacting and weathering where both metallic iron and iron silicates react with both oxygen and water.

Polymict Breccia: A breccia containing clasts derived from multiple sources. In polymict breccias found as meteorites, the clasts may be those derived from impact in the process of producing a planetary regolith.

Regolith: Loose, fragmental material on a planetary surface. On the earth, regolith is the inorganic component of soil—soil being unique to the earth with its biogenic (organic) component of humus.

Residence Time: The time that a meteorite (collected as a find) has remained on the earth's surface. Residence time can also refer to the time a meteor has remained in orbit around the sun—a time period capable of being determined from the effect of cosmic rays and the solar wind on components in the surface of the meteor.

Soil: Loose, friable material found on the surface of the earth. Soil contains humus, a complex organic material derived from the biosphere. As biospheres and biogenic material (probably) doesn't exist on other planets of the Solar System, humus is not found elsewhere, other than on the earth. Soil is therefore not found on any planetary body other than the earth.

Terrestralization: Chemical conversion of many of the components of stony meteorites to earth-like compounds by interaction with the earth's atmosphere and hydrosphere. With terrestrialization, the ferrous silicates become hydrous (ferric), iron oxide (limonite), and clay, metallic iron becomes hydrous ferric oxide and the nickel becomes greenish, soluble nickel compounds.

Terrestrial Residence Time: The time usually determined by the amount of terrestrialization which a meteorite has undergone.

Rhyolite: This wind abraded specimen came with a lot of NWA's. Rhyolite is an earthly felsic, igneous rock and is a rock representative of a highly differentiated planetary body, namely earth. Its red color excited a few persons, thinking that it might be a rock from Mars. Martian meteorites are grey or black, the red color of Mars is a surface weathering phenomena—the regolith of Mars is red. When a red Mars rock is broken open, it will be grey or black inside—or so we currently think!

Sliced and polished surface of the previously illustrated rhyolite meteor-wrong.

Chapter Seven

Rare (and not so rare) Terrestrial Nickel Minerals and Meteorites

Economic Nickel Deposits

In various parts of the world nickel deposits occur in quantities large enough for industrial mining. The most common nickel mineral in these deposits is pentlandite, an iron-nickel sulfide. Most of these nickel deposits occur in geologic terrains, which are very ancient—mostly originating from the Archean Era of geologic time, a time when the continents did not exist or were just beginning to form. This predominance of nickel deposits formed during the early earth lends support to the idea that the Earth, like the Moon and Mars, was subjected to huge impacts of asteroidal sized objects, many of which became incorporated into the hot, plastic crust of that time and the asteroidal material became remobilized as nickel deposits.

Millerite and Meteorites

Other than the nickel minerals found associated with ancient Archean rocks, nickel minerals generally are uncommon in earthly rocks. Archean nickel deposits generally consist of minerals like pentlandite, which, because it can occur in large quantities, can be mined on an industrial scale. Ancient well known mineable deposits are those of Sudbury, Ontario, and in the Canadian Midwest at Flin Flon,

Manitoba, both of possible asteroidal origin. Scattered over the globe, however, are also many small, localized nickel mineral occurrences that are suspect as having a possible extraterrestrial origin through meteorites. The author, being from the St. Louis, Missouri, area, has long been aware of the connection between nickel minerals and this region. One of the rare nickel minerals first discovered, the mineral **millerite**, is noted in late nineteenth century literature as occurring in the quarries of north St. Louis—St. Louis being one of the so-called Dana localities for this mineral.

These nickel minerals are peculiar to limestone beds of Paleozoic age, which occur extensively in the U. S. Midwest. They are distinguished by the fact that they **do not** occur associated with the sulfide mineral deposits of that region, which consist of lead, zinc, and copper sulfides and are generally known as MVT (Mississippi Valley Type deposits)—mineral deposits named after large lead and zinc sulfide deposits in the Mississippi Valley Region. Rather these nickel occurrences tend to be small and isolated from the MVT metallic sulfide concentrations. As such, they are not mined as ore, but are found, as are the MVT deposits, in undisturbed limestone and dolomite beds deposited on continental interiors in what is referred to by geologists as the craton. Because of the isolation of these nickel deposits, it has been proposed that the nickel in these occurrences has an extraterrestrial, meteorite related origin.

Rocks of the early earth: Geologic formations from the earliest period of the earth's history record evidence of extensive impact phenomena as well as containing nickel, chromium, and platinum deposits, material suggestive of being derived from asteroidal sources. These are 3.3 billion year old dark rocks from an Archean Greenstone belt (intruded by pink granite) which are part of the Superior Province of the vast Canadian shield.

Flin Flon, Manitoba, 1990: Another part of the large smelter in this northern Manitoba town based upon nickel mining.

Pentlandite, an iron-nickel sulfide: This nickel mineral is associated with very ancient parts of the earth's crust where it sometimes is mined. This nickel ore came from the Sudbury, Ontario, deposit of northeastern Ontario, Canada. Similar occurrences of pentlandite occur at Flin Flon, Manitoba, where it also forms the basis of extensive nickel mining operations.

Close-up of Flin Flon, Manitoba, smelter.

Nickel ore, Flin Flon area: The mineral forming the radiating crystals at the top is millerite, a mineral abundant enough locally in western Manitoba to mine as an industrial source of nickel.

This is part of a large nickel mining and smelting operation at Flin Flon, Manitoba, Canada. A similar operation is found at Sudbury Ontario, both in very ancient rocks of the ancient Canadian Shield. These rich nickel deposits have been hypothesized to have originated from asteroids that became incorporated into the crust of the early earth. The asteroids, being absorbed into the hot, plastic crust, were incorporated into it, forming the nickel deposits. The Sudbury, Ontario, deposit even includes shatter cones; however, these may have originated from a later impact.

Millerite associated with pyrrotite, Thompson, Manitoba. Millerite is the golden (yellowish) layer at the top with small radiating crystals. Millerite is nickel sulfide. The millerite is perched on top of a mixture of magnetite and pyrrotite (iron sulfide).

Looking down on a chunk of fibrous millerite crystals from Thompson Manitoba. The central and northern regions of Manitoba harbor concentrations of nickel minerals derived (possibly) from asteroidal origin during the earths early history (period of late heavy bombardment)..

Zaratite, Lord Brassey Mine, Heazelwood, western Tasmania. This rare nickel mineral is a hydrous nickel carbonate. It is a secondary nickel mineral derived from nickel sulfide associated with Precambrian untrabasic rock. Nickel minerals (some rare and found only at this locality) in these ancient geologic terrains, like that of Sudbury, Ontario, and Flin Flon, Manitoba, *may have been derived from asteroidal sources during the "late heavy bombardment" of the Moon and the early Earth.* (Value range F).

Fibrous layer of millerite crystals of the above photo viewed through a hand lens.

Zaratite: Another specimen of this unusual nickel mineral from western Tasmania.

Another view of fibrous millerite crystals from Thompson Manitoba as viewed through a hand lens

Zaratite, Tasmania—another view.

Meteorites

Meteorites come in two primary types (or categories), metallic meteorites and stony meteorites. Stony meteorites usually are made of a mix of silicate minerals (pyroxenes, olivine, and some feldspar) intermixed with (usually small) masses of a nickel-iron alloy. Many other types of stony meteorites also occur such as the SNC's or Martians (fragments of Mars blasted from the planets surface by impact), Lunars (fragments of the Moon (blasted from that body by impact), as well as other diverse types from asteroids and even some which may have come from comets. Most stony meteorites are similar to those illustrated here and their origin was primarily from various asteroids, mostly those occupying orbits between Mars and Jupiter. Many meteorites, the majority of them, contain nickel, an element which is normally rare in the earth's crust.

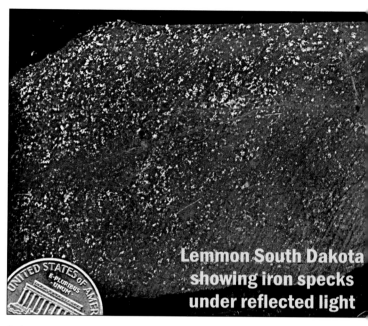

Lemmon South Dakota showing iron specks under reflected light

A slice through a stony meteorite: This is a find discovered in South Dakota near the town of Lemmon (for which it is named). The numerous white specks (actually metallic) are composed of an alloy of nickel-iron. (Value range E).

Find Gold Hill, Arizona

Meteorite, Gold Basin, Arizona: A large number of these stony meteorites have been collected from a dry region in Arizona. They are essentially identical to stony meteorites found in large quantities in the deserts of northwest Africa, the topic of Chapter Six of this book. It appears that a large number of these chondritic meteorites may have bombarded the earth a million or so years ago. Could the weathering and remobilization of nickel in these have been the source of the nickel minerals now found in vugs and other cavities as millerite?

Farmington, Kansas, fall 1892: A fireball accompanied by detonations came with this meteorite fall, which is possibly a fragment of an Apollo Asteroid (light grey areas are nickel-iron).

Fall Farmington, Kansas

NWA 2995
Achondrite, Lunar (ALUN)
Algeria
Found 2005
0.4754 gram

Lunar breccia: A lunar meteorite found in Algeria. Breccias such as this are produced by high velocity impact. This meteorite is composed of fragments generated on impact and later cemented together. The white specks are crystals of sodium feldspar. Lunar meteorites are being found worldwide more frequently all the time, but they still are rare so that they command high prices among collectors. They could not be a source of sporadic nickel minerals because of (1) their rarity and (2) lunar meteorites are very low in nickel.

Canyon Diablo, Arizona
Widmanstatten structure
has been partially
obliterated by shock

St. Francois County, Missouri: A slice of a nickel-iron meteorite (siderite) from a mass found in the same general area where millerite (nickel sulfide) is found. Such an association is probably coincidental as the nickel in the millerite probably came from a meteorite source that fell millions of years ago. A meteorite like this, although more resistant to weathering than a stony one in a humid climate, would still disintegrate by weathering in a few thousand years. If it had done so, however, its 10 percent nickel content could be a source for nickel, which, if carried downward by groundwater, could form millerite in a suitable geologic environment.

Canyon Diablo, Arizona: A nickel-iron meteorite of which thousands of specimens have been collected. They are associated with a large and fresh impact crater (Meteor Crater) in Coconino County, Arizona. Like the strewn field which produces the Gibeon meteorites, enough metallic meteorite fragments occur (or occurred) at this locality that, upon weathering, could have produced secondary nickel minerals in some of the underlying rock layers.

Gibeon
Fine Octahedrite
Find

Meteor crater
Winslow, Arizona

Gibeon, Namibia: A slice through a finely crystalline nickel-iron meteorite (fine octahedrite). Gibeon has been widely distributed among museums and collectors. It is found over a strewn field which has supplied numerous specimens to collectors. In the strewn field where it is found, nickel would occur in large enough amounts to form, upon complete weathering, a localized nickel deposit if the nickel were to be remobilized on weathering.

Aerial view of Meteor Crater, a relative fresh earthly impact site in northern Arizona, the source of the Canyon Diablo meteorites.

Millerite and its Alteration Products: Pecoraite and Honessite

In contrast to desert regions where meteorites can survive on the surface of the earth for short periods of geologic time, in humid regions like the US Midwest, they quickly weather away, ending up as masses of iron-rich clay. A major component of most meteorites is nickel, an element that otherwise is rare in rocks of the earth's crust. Unlike other sulfide mineral deposits in many parts of the world, these Midwest millerite occurrences and its alteration products have no obvious relationship to any hydrothermal activity or deep seated igneous source. Most of the earth's nickel has been "locked up" in the core during formation of the planet and is inaccessible. Millerite and its alteration products, Pecoraite and Honessite are localized in their occurrences in a manner suggestive of the strewn fields of stony meteorites. It has been suggested that these localized occurrences of nickel originated from meteorite falls of the past in which the nickel from a strewn field of meteorites went into solution from weathering and then was transported downward by descending groundwater precipitating as the sulfide in favorable horizons of underlying limestone and dolomite. These favorable horizons being ones which had sufficient porosity and permeability and could combine the relatively soluble nickel with sulfur which was already present in the limestone.

Chert (silica) nodule with millerite in cavity to right of penny. Arnold, Jefferson County, Missouri.

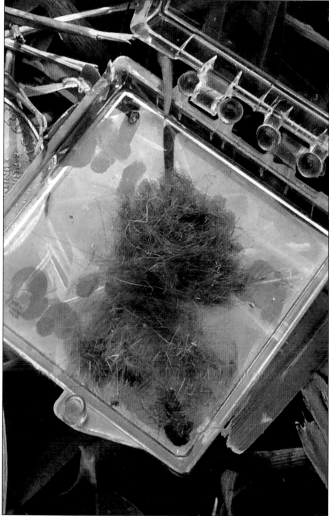

Close-up of millerite "steel wool" mass from a chert nodule occurrence south of St. Louis (Arnold, Missouri). These various forms of millerite represent a rare and unusual mineral. Good specimens of millerite are desirable. (Value range F).

Millerite from cavity in a chert mass from the same locality as above photo. (Value range F).

Another chert chunk containing cavities (vugs) with milerite. Small needles of millerite are in a vug to the right of the penny. *Courtesy of Brad Weatherbee*

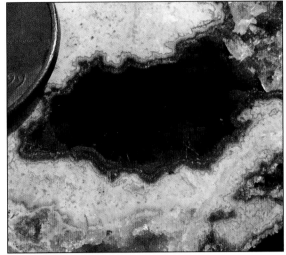

Close-up of same black quartz lined vug with small millerite crystals.

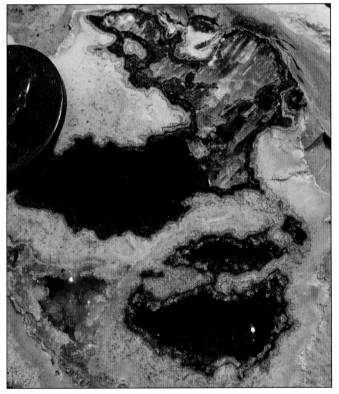

The same vug as in previous photo. Note that a portion of the vug has been filled with calcite, Arnold, Missouri.

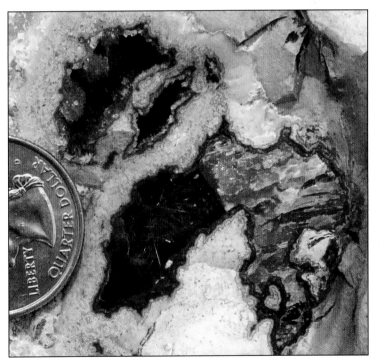

Flip side of the previous vug with numerous fine millerite crystals.

The reason that millerite is found separate from other sulfide mineral occurrences may be from the source of the nickel being different from that of more common MVT minerals, which derive their lead, zinc, and copper from small amounts of these elements deposited when the limestone or dolomites were formed in shallow seas. The occurrence of millerite, sometimes found within calcite vugs and geodes, is also suggestive of a source different from that of other sulfides, which can occur in a similar manner but **never occur** with the nickel minerals (but see disclaimer).

Nickel is one of the elements which has been sequestered in the earth's core, most rocks and ore deposits lacking the element. Meteorites, on the other hand, except for a few rare types, are well endowed with nickel; in some of them nickel is the second most abundant element in them, as is the case with nickel-iron meteorites or siderites. It is of interest that nickel is almost totally absent from MVT mineral occurrences worldwide; the large deposits mined for decades near Joplin, Missouri, and the Old and New Lead belts of the Ozarks lack nickel minerals, but do have cobalt, an element present in meteorites but occurring in them in nowhere near the abundance of nickel. Being a sulfide mineral, millerite would be expected in large ore bodies like that of Joplin, Missouri, or the Missouri lead belts, but it **doesn't occur there**; nickel mineral occurrences in limestones and dolomites (the setting of MVT minerals in the above ore bodies) **are always isolated** from other metallic sulfide deposits. Such isolated occurrences have led to the hypothesis that nickel minerals like millerite originated from meteoritic sources. This extraterrestrial origin was not taken seriously in the past, however, as more connections between terrestrial geology and extraterrestrial phenomena are made, an extraterrestrial origin for these nickel minerals is no longer so "far out."

Dolomite vugs like this occur in a layer of Cambrian age strata in Missouri's Old Lead Belt and can locally contain spectacular millerite specimens. These zones are geographically and stratigraphically isolated from strata which has yielded vast amounts of galena (lead sulfide). The orange mineral is ferroan dolomite; a form of the mineral dolomite containing small amounts of iron substituting for magnesium atoms in the crystalline structure.

Vug of ferroan dolomite crystals with a small spray of millerite. Old lead belt area (Park Hills), St. Francois Co., Missouri. (Value range D).

Geode interior with steel-wool millerite from Halls Gap, Kentucky. The Halls Gap locality is one where nickel minerals occur; possibly sometime in the past it was a strewn field of meteorites that were totally weathered away, with nickel remobilized as millerite deposited in the cavity of the geodes. *Courtesy of Joseph Lobecz.* (Value range F).

Millerite needles. Possibly one of the most spectacular sprays of this mineral ever found—this specimen graced the cover of the 1997 Missouri issue of *Rocks and Minerals* magazine. Old lead belt region, St. Francois Co, Missouri. *Courtesy of Glenn Williams.* (Value range A).

Millerite in vugs or geodes can be found embedded in calcite, as seen in this reflected light view of a specimen from southern Lincoln County, Missouri. (Value range E).

Another view of the above millerite specimen.

Small Geode with enclosed millerite, Biggsville, Illinois. (Value range F).

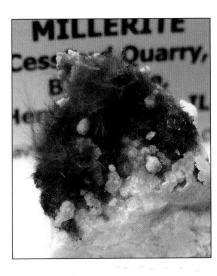

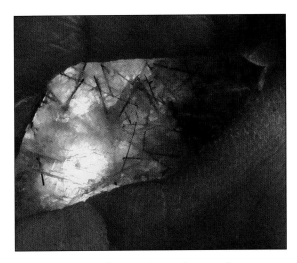

Calcite containing millerite in cleavage fragment from near Troy, Missouri, a region that locally has produced various forms of millerite. The region where this came may represent another original stewn field that is now a totally weathered site.

More millerite sprays in chert.

A nice group of millerite occupying tight cracks in chert, Lincoln County, Missouri. (Value range F).

Another calcite cleavage fragment with millerite needles. Delicate crystals like these millerite needles embedded in another transparent mineral make attractive and interesting mineral specimens. Also the millerite, unlike that occurring as exposed needles and steel wool-like masses, is protected by being sealed in the calcite. Needles of the mineral rutile (titanium oxide) are sometimes found embedded in clear quartz, which is associated with pegmatites. Such quartz is called rutilated quartz. Here is "millerated calcite!" (Value range E).

Millerite in calcite from another locality in the Troy, Missouri, area.

Millerite in chert, Troy, Missouri. Cherty limestone of the Lower Mississippian (Kinderhook age) in a quarry north of Troy carries these sprays of Millerite. The region of northern St. Charles and southern Lincoln counties of Missouri appear to have localized occurrences of rare nickel minerals. (Value range G, single specimen).

Another group of millerite crystals which occupy slight cracks in the chert. Troy, Missouri. (Value range G).

Millerite crystals in calcite vug, Rock Hill quarry, St. Louis Co., Missouri. Joe Schraut collection. *Courtesy of Joseph Lobecz.*

Another view of the same specimen as shown above.

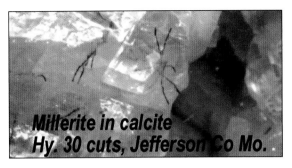

Small millerite crystals in calcite, Jefferson County, Missouri. (Value range G).

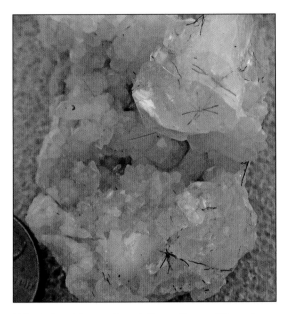

Millerite in calcite, northern Jefferson County, Missouri.

Millerite? There is some suggestion that these slightly bent needles are marcasite crystals rather than millerite, which they resemble. They come from vugs in Ordovician limestone—most of the millerite from eastern Missouri comes from Mississippian limestones.

Another view of the above specimen.

Pecoriate: A concentration of this nickel sulfate mineral was found during construction of I-44 and Antire Road in western St. Louis County in 1962. The pecoriate occurred inside geodes as alteration products of clustered millerite needles. Some sources give this occurrence as being from Antire Road Quarry; however, this abandoned quarry in Ordovician limestone has not produced nickel minerals as far as is known by the author. The pecoriate bearing geodes also occur in the Fern Glen Formation and not in the Salem and St. Louis limestones as stated in some sources. These green pecoriate crystals are pseudomorphs after millerite. (Value range E).

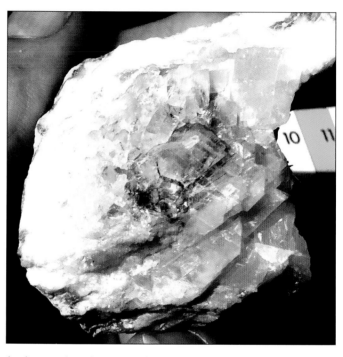

A calcite geode with pecoraite (nickel sulfate) intermixed with the calcite, Antire Road, I-44 locality St. Louis Co., Mo.

Pecoraite in quartz (geode) I44 and Antire Road St. Louis Co., Missouri

Pecoriate containing geode from Antire Road and I-44. (Value range F).

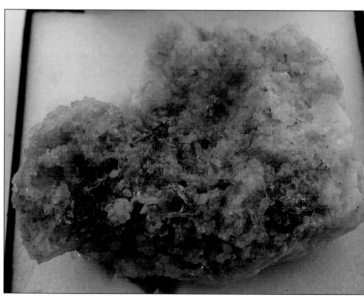

Calcite (with some quartz) geode with pecoraite—some being pseudomorphs after millerite.

Pecoraite in quartz geode

Millerite needle mass altered to pecoraite, Antire Road-I-44 locality. (Value range F).

Honessite: Sprays of this rare nickel mineral were collected by the author from construction of a lane of highway 61 north of Troy, Missouri, in 1976. (Value range E).

Another specimen of Honessite from north of Troy, Missouri. (Value range D).

Bibliography

Brotherton, Baley, 1948. "Millerite in Iowa." *The Mineralogist*, Nov. 1948. p. 530.

Nininger, Harvey H., 1974. "A Case for Impact Geology." *Earth Science*, Vol. 27, No. 5. (Sept., Oct.).

_____.1974. "Those 'Earth Grazing' Asteroids." *Earth Science*, Vol. 27, No. 1 (Jan., Feb.).

Sinotte, Stephen R. 1953. *The Fabulous Keokuk Geodes. Geodes*. Portage, Michigan: Self Published.

Glossary

Astroblem: An impact structure on the earth's surface that has been severely modified by erosion and weathering so that it no longer is a crater; however, the effects of high velocity impact on rocks are still evident.

Chert: A chemically precipitated sedimentary rock composed of quartz. Chert is a common rock on the earth. It is not present in meteorites, as is also the case with other forms of quartz.

Geode: Globular-shaped rocks composed of quartz, often which are hollow and contain crystals. Millerite or alteration products of millerite are sometimes found in the interior of geodes.

MVT (Mississippi Valley Type mineralization): Occurrences of sulfide minerals of lead (galena), zinc (sphalerite), as well as other, more rare elements like cobalt in beds of limestone and dolomite in the US Midwest. The strata in which these mineral deposits occur show no evidence of mineralization related to any igneous activity or to hydrothermal activity—a characteristic of many sulfide mineral deposits. Nickel minerals are notoriously absent from MVT type occurrences, themselves being rather found separate from them in similar carbonate rocks (limestone and dolomite) in isolated occurrences.

Old and new lead belts of Missouri: Large MVT type deposits of lead sulfide (galena), which occur in the Ozark Uplift of Missouri. The old lead belt in St. Francois County was mined from the late nineteenth century until 1970; the new lead belt, 50 miles to the southwest, is still in production. Nickel minerals have been found in the vicinity of both of these "lead belts," but never actually associated with the actual mines or ores.

Chapter Eight
Meteoritical Odds and Ends

Astroblems
of the Ozark Region

Ancient parts of the earth's surface show evidence of large impact scars. One of these geologically old parts of the earth is the Ozark Uplift of Missouri and Arkansas. These impact scars represent ancient impact craters worn down from erosion over millions of years and are known as **astroblems**. Associated with them are breccias (highly fractured and fragmented rock) sometimes associated with pulverized rock known as rock flour. Also associated with astroblems are peculiar cone-shaped structures known as shatter cones and shocked quartz—the latter a form of the common mineral quartz, which, as a consequence of high velocity impact, has had its crystalline structure changed. Astroblems scattered over the earth indicate that the earth, like other planets of the inner Solar System, had a history of impact phenomena similar to that of Mars and the Moon. Similar, but without the high crater density of these "worlds," as much of the cratered terrain of Mars and Luna was produced during the first billion years of their history, a record from a period of geologic time almost totally absent from the earth. Over 170 astroblems have currently been located on the earth's surface. They are characterized by circular structures made up of highly fractured rock (breccia), with an uplifted central area associated with a form of quartz known as "shocked quartz" or coesite, a polymorph of quartz. Such ancient impact structures during the past thirty years have been found to be more common than previously recognized. Phenomena associated with many of these structures, if it is distinctive and available, can be collectible; although, impact related phenomena like shatter cones and impact breccia are not as desirable and collectible as are actual meteorites.

Meteor Crater in northern Arizona is some 40,000 years old—it is geologically young. Over millions of years this crater will be removed by weathering and erosion, but the underlying fractured and brecciated rocks will still be present. The impact crater would now have become an astroblem!

Crooked Creek Structure

An asteroid impacted southern Missouri some 230 million years ago where now flows a major tributary of the Meramec River known as Crooked Creek. Here impact produced a crater at least six miles in diameter, an impact site considerably larger than the Winslow Crater of Arizona; however, the crater itself has long been removed by millions of years of erosion. What is seen today is the contorted and "broken" geology produced by impact on the underlying rock strata. The Crooked Creek Structure on a geologic map resembles a large bullseye where, at the center of the bullseye, the rocks are extremely brecciated and shatter cones are present. The Crooked Creek astroblem is one of a number of such structures occurring in the Ozark region of Missouri and Arkansas, a part of the earth's surface that is fairly ancient and which has not been covered by glaciers or other geologically recent events that might otherwise conceal such impact scars. Other Ozark astroblems widely recognized by geologists are the Decaturville Structure and the more recently accepted Weaubleau Creek Structure.

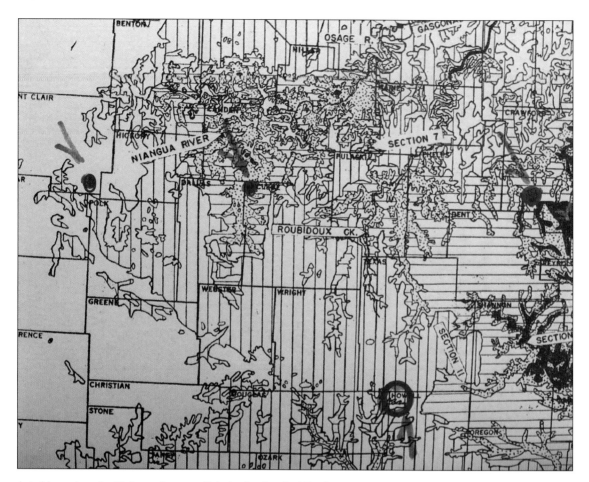

Astroblems along the 38 degree linament. Right (red)—Crooked Creek Structure. Middle—Decaturville Structure. Left (blue)—Weaubleau-Osceola Structure (blue). The Douglas County, Missouri, knobs are shown (Blue circle) at the bottom. They are in both Douglas and Howell Counties, mostly in Douglas. These sites are superimposed on a geologic map of part of southern Missouri.

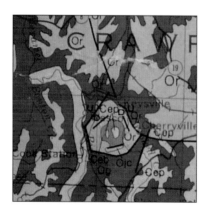

The Crooked Creek astroblem in southern Missouri as shown on the 1979 version of the Missouri Geologic map. What was once a 7-8 mile diameter crater (much larger than Meteor Crater shown above) is now represented by highly disturbed geology, the actual crater having been removed millions of years ago by erosion. Nickel minerals have not, to the authors knowledge, been found associated with this structure. Their absence was (or is) one of the reasons that some geologists originally questioned its extraterrestrial origin.

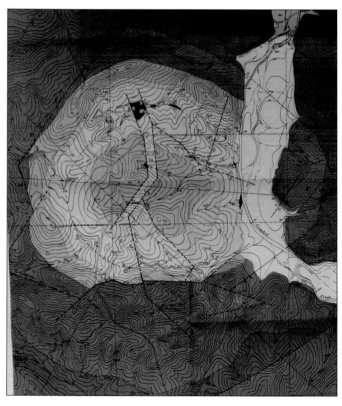

More detailed geologic map of the Crooked Creek Structure.

Impact breccia, Crooked Creek structure. Limonite forming a matrix between chert clasts gives a good contrast between the rock (chert) broken on impact (the clasts) and material filling between them, which originally was rock flour. (Value range F for similar breccia).

Another group of shatter cones from the Crooked Creek structure.

Impact breccia from the Crooked Creek structure cemented into a masonry wall. Boulders of brecciated chert such as this occur widely over the Crooked Creek astroblem. In fact, all of the rocks associated with this astroblem are extensively brecciated. (Value range G for similar breccia).

Brecciated fossil snails: These fossils came from the highly brecciated rocks of the Crooked Creek Structure. They are from the Lower Ordovician, some 490 million years ago and have been brecciated along with the associated rock. Small iron oxide masses on the chunk conceivably could have come from some of the iron from the impacting asteroid, which hit some 200 million years after the snails were living in what was then a shallow sea. (Value range E).

Shatter cone from the Crooked Creek structure. Extraterrestrial impact generates, from high velocity shock waves, a series of cone-like structures like these, which have a diagnostic signature. Such a signature is produced only by high velocity events, such as that which produces an impact crater. Presence of shatter cones is one of the criteria necessary for a geologic structure like the Crooked Creek structure to be considered as an astroblem of extraterrestrial origin. (Value range F, single cone)

The back side of the same specimen as shown above. The dark masses are pseudomorphs of pyrite, an iron mineral. Iron minerals occur throughout the Crooked Creek astroblem. It is possible that some of this iron may have been derived from the impacting asteroid. Minerals associated with large impacts like this often migrate after impact. Only if nickel were present might the source of the iron be nailed down as being of extraterrestrial origin. Nickel compounds are often removed from such structures because of their solubility in groundwater.

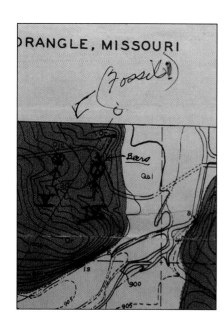

Part of the Crooked Creek Geologic map showing the tilted and brecciated Jefferson City Formation from which the above fossil snails came.

Another group of fragmental trilobites collected from crystalline limestone and dolomite near ground zero. The trilobites consist of cephalons (heads) of the lower Upper Cambrian genus *Tricrepacephalus texanus*. (Value range F).

Trilobite fragments from Cambrian strata brought up by an impact taking place near ground zero. Ground zero is the center of the structures where the impacting asteroid made contact with the earth.

Peculiar snail-like (left hand coiling) mollusks (*Scaevogyra swezeyi*) that (presumably) were fractured in the process of being brought to the surface by rebound of the impact. Were it not for the impact, strata yielding this fossil (as well as the above trilobites) would still be buried well below the surface in this part of Missouri. (Value range F).

Decaturville Structure

The Decaturville Structure resembles the Crooked Creek structure in most of its characteristics. One of the peculiarities unique to it is the presence in its center (ground zero) of masses of coarsely crystalline igneous rock (pegmatite) and metamorphic rock (mica schist), rocks otherwise rare in this part of North America. These unique (for the area) rocks appear to have been brought up to the present surface from at least 1,200 feet below by impact rebound, the same phenomena that produces the central spires of some lunar and other large planetary craters.

Very coarsely crystalline granite-like rock (pegmatite) from ground zero of the Decaturville astroblem. This rock was brought up from ancient "basement" rock some 1,200 feet below the present surface by impact rebound. Such rebound in Lunar and Mercurian craters has produced central spires. To produce such rebound takes tremendous amounts of energy—enough energy to launch earth rocks into space. Possibly some Missouri rocks are on the Moon or on Venus or Mercury and were launched from this impact. The rock to the right is mica schist, also a rare rock from Missouri brought up by impact rebound.

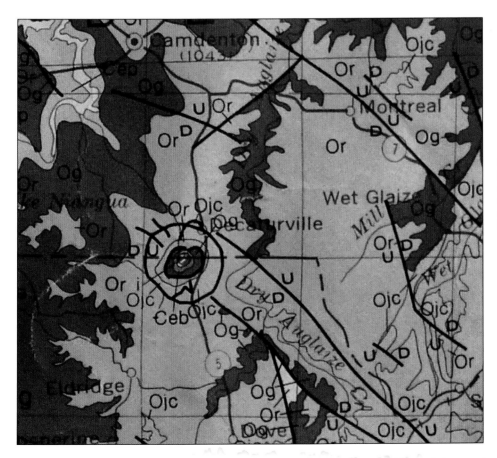

Geologic map showing the Decaturville Structure, located south of Lake of the Ozarks. Ground zero of the structure is the small red center, an area of basement rocks brought up by impact rebound.

Lunar crater with central spires created by impact rebound, center of photo. (*Courtesy of NASA*).

Impact breccia from near ground zero of the Decaturville Structure. Angular clasts, similar to those of the Crooked Creek Structures are present. Breccias are very commonly associated with extraterrestrial phenomena.

Weaubleau-Osceola Structure

The third large impact structure in the Ozark region is the Weaubleau-Osceola Structure located west of the Decaturville Structure. Its geology is somewhat more confusing and less clearly impact-related than the Crooked Creek and Decaturville structures and for this reason until recently it was not considered an astroblem. It is of interest that all of these, the Crooked Creek, Decaturville, and Weaubleau Creek structures occur in a line located at approximately 38 degrees latitude (the 38th parallel lineament). The Weaubleau Creek Structure was the last of three problematic geologic structures on the liniment to be confirmed as such. An impact origin for this peculiar structure was encouraged by some interesting drilling and the acquisition of core samples that exhibited shatter cones, brecciation, and peculiar shale intrusions. Aside from drilling, there is the occurrence of fractured and jumbled rock strata well exposed along Weaubleau Creek, a tributary of the Osage River. Another oddity probably related to the structure is an abundance of chert spheres at what appears to be the center of the impact structure (ground zero). These rock balls are used locally in producing imaginative masonry work. They are known as Weaubleau eggs and are hypothesized to have formed through some sort of nucleation process triggered when the impacting asteroid fell into a shallow sea during the Mississippian Period, some 330 million years ago. Other possible impact-related structures occur throughout the Ozarks of Missouri and Arkansas; however, some of these may be from land use patterns or they may be sink hole related. Associated with some of them are breccias which conceivably may have been formed from impact. The critical occurrence of shatter cones and shocked quartz however has not been found in any of them (so far).

Right:
Peculiar spherical chert concretions known as Weaubleau "Eggs," which occur near ground zero of the Weaubleau-Osceola Structure. (Value range G, single "egg").

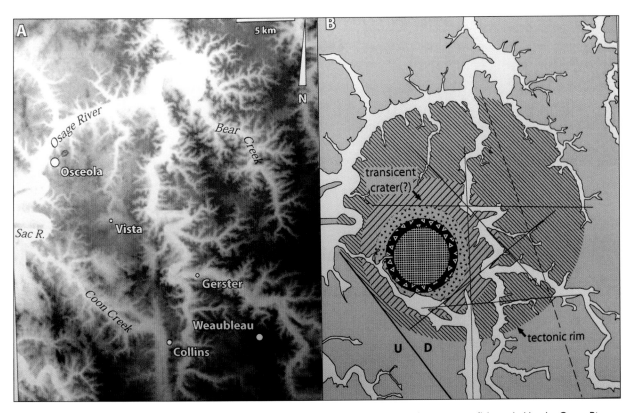

Weaubleau-Osceola Structure, Missouri. Left: Shuttle radar image showing circular feature (impact crater?) bounded by the Osage River to the northwest and Bear Creek to the northeast. A circular pattern can be seen north of Vista. Right: Inferred crater deformation derived from radar image, surface geology and topography. From Evans, K., Miller, J. and G. Davis—*Field Trip Guide to Geology of Weaubleau-Osceola Structure, Southwestern Missouri*, 2005.

Another group of Weaubleau "eggs."

Douglas County, Missouri, Knobs and Breccia Boulders

A cluster of puzzling knobs occur in the south central portion of the Missouri Ozarks, primarily in Douglas County. These knobs are capped with huge boulders of what resembles impact breccia. The definitive astroblem phenomena of shatter cones and shocked quartz has not been found in association with these knobs, but a type of rock flour cements the angular clasts together that make up the large boulders—large boulders which cap a cluster of five or more knobs. The author has also found small geodes containing the green nickel mineral pecoraite discussed in Chapter Seven just south of one of these knobs. The following schematics show a mechanism as to how the siliceous breccias capping the knobs may have formed as well as how the knobs themselves may have formed.

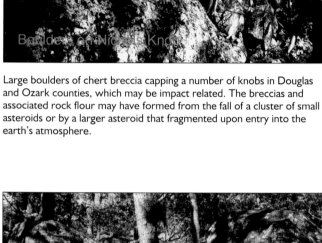

Large boulders of chert breccia capping a number of knobs in Douglas and Ozark counties, which may be impact related. The breccias and associated rock flour may have formed from the fall of a cluster of small asteroids or by a larger asteroid that fragmented upon entry into the earth's atmosphere.

Breccia capped knob (Nichols Knob)

Breccia capped Nicholas Knob, Douglas County, Missouri. Clustered together in the central Ozarks of southern Missouri are found a group of breccia covered hills that may have an impact related origin. Each knob is capped by large boulders of breccia having features suggesting its formation from high velocity impact. Shatter cones have not been found with the breccias, but small geodes containing the nickel mineral pecoraite have been found nearby.

Nichols Knob

Boulders of impact(?), breccia capping Nichols Knob. These breccia boulders are localized on the knobs of Douglas County and have a signature which is distinct from other chert breccias of the Ozarks—breccias which appear to be related to silcretes.

Impact(?) breccia boulders capping Blue Buck Knob.

Brecciated Boone chert, Blue Buck Knob

Close up of sliced breccia from Nichols Knob. The clasts are composed of chert cemented together by finely pulverized chert (rock flour).

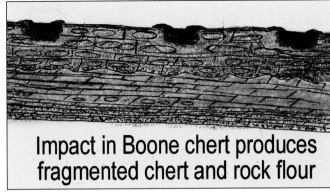

3.

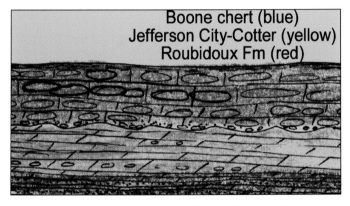

Mechanisms suggesting how breccias on Douglas County knobs of southern Missouri may have formed.

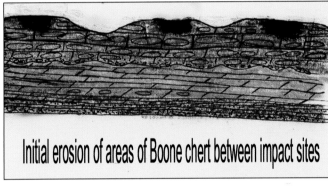

4. Stream erosion cuts into the Boone chert, which also consists of such limestone, which also dissolved away.

1.

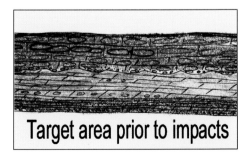

2.

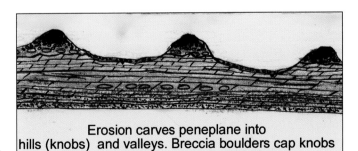

5. The final result is that only remnants of the Boone chert (blue) remain. These remnants are the large boulders of chert breccia, which cap the knobs.

Steinheim Astroblem

A unique astroblem occurs in the southern German Province of Bavaria near the town of Steinheim. It is unique in the fact that the impact crater, after being formed upon impact, became a fresh water lake, the limy sediments of which, being derived from impacted Jurassic limestone, preserved a variety of life forms as fossils. The actual impact that produced the crater happened approximately 25 million years ago during the Miocene Epoch of the Cenozoic Era of geologic time. Fossils found in the fresh water limestone that formed in the crater consist of mammals, birds, and reptiles, as well as the more commonly found fish and plants.

Meteorite Ownership

Exactly who owns a meteorite when it is either found (occurs as a find) or is collected after a fall is potentially a "sticky wicket." In many (most?) cases, it's "finders-keepers;" however in the US, technical legal ownership of a meteorite is with the person or institution who owns the land upon which the meteorite fell or was found.

In different parts of the globe claims range from total ownership by the government or the national museum of **all** meteorites to the more laid back "finders-keepers" attitude. The range of interpretations regarding this matter is varied and often is inconsistent from one occurrence to another—often dependent upon the attitudes of local authorities and their personal views. This matter also involves the issue of science and education. Regarding science, it might appear that science would benefit most if **all** meteorites went into academic or governmental institutions like museums. This attitude, however, can be counterproductive in that it discourages collectors as well as "poisoning" the collector-scientist relationship. With meteorites, as with fossils, it has been said that the more eyes looking for them, the better for all. Most meteorites, whether they are falls or finds, usually require public participation in some manner for their retrieval. An exclusive attitude is also not warranted in that not all meteorites (whether they be finds or falls) have scientific value or at least significant scientific value—after all, how many specimens are really needed from the same find or fall?

The author has been questioned as to the "correctness" of meteorite collecting and ownership in general, the questioners suggesting that ownership is not a necessary prerequisite for interest in meteorites (or fossils for that matter) and that personal collecting interferes with the interests of serious science. Regarding this mindset, my rebuttal is that collecting and ownership is a part of human nature! With regard to meteorites (as well as fossils, minerals, artifacts, coins as well as other collectibles), great expertise regarding such items often comes as a consequence of collecting and ownership. With meteorites, Harvey H. Nininger comes to mind; he put together, because of intense interest in them, one of the largest meteorite collections ever assembled and contributed ideas on meteorites that were well ahead of his time. There is no better way to become acquainted with meteorites than to work with them firsthand. It's been said that "the best geologist is the one who has seen the most rocks" and this dictum applies equally well to meteorites. A small percentage of persons will really be fascinated by them and working with them either through possession or through the availability of a collection is seen as the best way of acquiring familiarity with meteorites. Such familiarity often starts with a persons acquiring a few specimens. It is for this reason that the author is so favorably disposed toward the bonanza of NWA meteorites. They make meteorites accessible to a greater number of persons. Artificially curtailing meteorite availability curtails what is an effective educational tool and interferes with the hands-on benefits which accrue with meteorite collection and ownership. The availability of NWA meteorites has increased meteorite availability. This should ultimately create more interest in meteorites overall, and this, in the author's eye, is a desirable outcome.

Steinheim Astroblem: Areal view of a well-defined astroblem. The central uplift (rebound) area is also well defined. Miocene lake sediments constitute flat terrain devoted to agriculture.

Shatter cones. Shattercones from Steinheim developed in a fine grained Jurassic limestone are especially distinctive. (Value range E).

Fossil fish (*Tinca* sp.) from Miocene lake sediments that filled in the impact crater. (Value range F).

Monomict breccia: With both clasts and matrix-between-clasts made up of tan Jurassic limestone—the bedrock of the Steinheim impact site.

Fossil reed impression ("large grass") from Steinheim crater lake. Reeds are large grasses that appeared on the earth only a few million years before formation of the Steinheim impact site. (Value range E).

Freshwater snails (gastropods) of the genus *Planorbis*, which lived in prodigious numbers in the lake of the Steinheim impact site. (Value range F).

Slab of polymict breccia from the Vredefort Ring or Dome, South Africa. Astroblems are found worldwide! This is an impact breccia from one of the largest observable impact sites on earth. Vredefort Dome was one of the first astroblems to be recognized as such and its recognition as an impact site by geologists (who before the 1960s generally refused to acknowledge such phenomena) was favorably commented on by H. H. Nininger. The breccia slab shown has numerous angular clasts or different types (it's a polymict breccia) scattered through a matrix of what was originally rock flour generated from the high velocity impact. (Value range F).

K/T Boundary and Extinctions, the Terminal Mesozoic Extinction Event

Dinosaurs fascinate the public and their extinction was accompanied by the extinction of many less-well-known life forms as well. These life forms, as is the case with dinosaurs, are known only from the fossil record. The extinction event of 67 million years ago at the end of the Mesozoic Era was one of the largest and most profound of the entire fossil record—it has been a mystery ever since the fossil record has been objectively studied! One hypothesis floating around since the 1960s was that these extinctions were caused by complications resulting from a supernova which took place somewhat near the Solar System 67 million years ago. Searching across the K/T boundary (Cretaceous-Tertiary Boundary) for trace elements that might represent decay products derived from short-lived radioisotopes, which accompany such a supernova, a spike (higher than normal trace element concentration) of the element iridium was found at this boundary. Iridium is a siderophile, platinum group element concentrated in nickel-iron meteorites—on earth most iridium is locked up in the planet's core. The finding of a worldwide concentration of iridium at the extinction point at the end of the Cretaceous Period was hypothesized as coming from an asteroid that struck the Earth 67 million years ago. Such a massive impact would have created a series of ecological disasters capable of causing mass extinctions like those seen at the K/T boundary. Such an event of this magnitude would have also broadcast vaporized meteoritic material worldwide to produce the iridium spike.

Direct evidence of such an impact was connected, a decade ago, to some catastrophic geology previously discovered in subsurface drilling for petroleum by Mexico's Pemex Company in the Yucatan region of Mexico. Various additional discoveries recently have made it even more convincing that such a giant impact took place in the southwestern portion of the Gulf of Mexico and that this impact was from the earth being a target for a large asteroid.

Impact? Breccia cobble from an Archean Greenstone belt of northern Minnesota cemented into a rock wall. Some of the earth's oldest events and the earliest geological events they record are found in the vast region known as the Canadian Shield, a region that makes up over a third of the North American continent. Breccias such as this, somewhat common in these ancient rocks, may have originated from extraterrestrial impact on the early earth. The early earth appears to have been extensively bombarded by meteors, the intensity of which diminished with the passage of geologic time.

K/T boundary, Wyoming: The K/T boundary is marked in the road cut by a change or absence of fossils below and above the boundary. Cretaceous strata in the road cut contain ammonites and in non-marine layers can contain dinosaur bones. Overlying Tertiary layers above the word "boundary" lack both of these critical fossils but contain fossils of organisms similar in many ways to those living today. This abrupt disappearance of many life forms, discovered in the mid-nineteenth century, formed the basis for delineation of the end of the Mesozoic Era—the middle-life era of the geologic time scale, which was being formulated at that time.

K/T boundary, southeastern Missouri: The K/T boundary near what was the head of an embayment of the Gulf of Mexico sixty-seven million years ago. The brown (bottom) layers contain an abundance of Late Mesozoic (Cretaceous) fossils—the greenish layers above it (Clayton Formation) just above the head of the boy on the left, does also, but its fossils appear to have been washed in rapidly from what is hypothesized to have been a giant tsunami generated from an asteroid impact in the southwestern portion of the Gulf of Mexico. Greenish clay of the Clayton Formation also contains volcanic ash (bentonite) and its overlying layers may represent sediments from land areas stripped of vegetation.

K/T boundary, Ardeola, Missouri: This is an exposure (outcrop) of the Clayton Formation, a sequence of strata that (probably) contains the iridium spike that ended the Mesozoic Era with the extinction of dinosaurs, ammonites, coiled oysters, and many other life forms typical of the Mesozoic Era. The lower beds of greenish clay contain Cretaceous fossils; the upper beds, those of the early-most Cenozoic Era. The K/T extinction event is at the middle of the greenish clay bed.

Collecting marine fossils from just below the K/T boundary from what was once the Gulf of Mexico at the end of the Mesozoic Era. Ardeola, Missouri.

Fossil zone—just below the K/T boundary.

High-spired snails (*Turritella* sp) collected from the previously shown outcrop. These marine snails were especially abundant in the Gulf of Mexico during the time of the Mesozoic extinction event.

Exogyra sp.: These large, coiled oysters went extinct at the K/T boundary; they, like the above snails, were especially abundant in shallow waterways of the Gulf of Mexico.

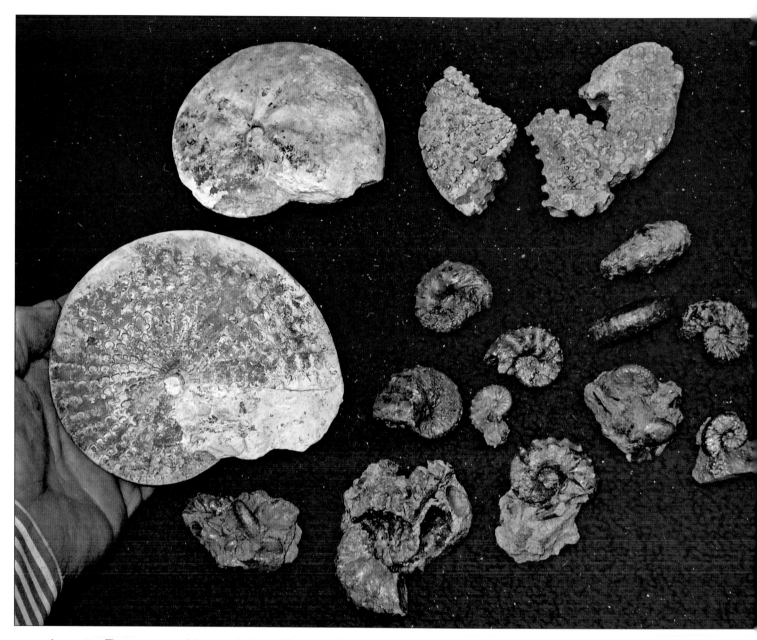

Ammonites: These were one of the major fatalities of the terminal Mesozoic Era extinctions. The modern octopus is believed by many paleontologists to be a shell-less ammonite that somehow survived the K/T extinctions. These were collected from outcrops at the Ardeola, Missouri, site shown above.

Right:
More Late Cretaceous extinction fossils from various localities world wide. Left-ammonites from Madagascar. Middle-ammonites from Europe. Right-dinosaur and mosasaur teeth and from Morocco.

K/T boundary extinction fossils: Fossils from outcrops below the K/T extinction event boundary include ammonites (left), coiled oysters (middle), and mosasaur & dinosaur vertebrae (right).

A Few More Meteoritic Odds and Ends

In the summer of 1964 the author engaged in summer geologic mapping in the western Ozarks of Missouri in Hickory and Benton Counties. In the process of this work, numerous land owners were met and interesting discussions often resulted—land owners who at the time were typical rural Ozarkians, friendly persons who generally were aware of and interested in nature and natural phenomena. I might add, they were also interested in what a geologist had to say about their land. In addition to the usual discussions on geology, rocks, and fossils, occasionally the topic would turn to meteorites, of which I've always had an interest. Sometimes a suspect "meteorite" would be brought out and these usually turned out to be chunks of limonite, goethite or hematite, iron minerals that occur ubiquitously over the Ozark landscape. In one case, however, an account had all of the characteristics of a presumed fall and the witness giving the account had the evidence to back it up: evidence which had been placed in a jar and stored in a barn. The jar, holding its remains, was brought out—appropriately labeled—and the account really appeared to have some merit. This included a retort followed by a whizzing sound in an early Spring evening of 1951. Neighbors also reported hearing the same sound and one of them picked up what appeared to be a chunk of black ice. This was placed into a container where, over a period of a few hours, it melted and turned into what appeared to be bad smelling water. When this water evaporated, a grayish residue remained which was then placed in the above mentioned jar. The residue resembled grey bird droppings and was crumbly and ugly looking. The question is, was this an ice meteorite, or possibly a fragment of a comet? The phenomena of ice falling from the sky or dirty ice melting, evaporating, and leaving behind a residue has been reported elsewhere but to date no "ice meteorite" has actually been collected.

In the summer of 1964 the author engaged in summer geologic mapping in the western Ozarks of Missouri in Hickory and Benton counties. In the process, numerous land owners were met—most of whom at the time were typical rural Ozarkians, friendly persons who were aware of and interested in nature and natural phenomena. In addition to the usual discussions on geology, rocks, and fossils, occasionally the topic would turn to meteorites. "Meteorite finds" usually turned out to be chunks of limonite or hematite. In one instance, however, an account of a presumed fall had some apparent merit. The related account included a retort followed by a whizzing sound in the early evening. Neighbors also reported this same phenomena and one of them picked up a freshly landed mass of what appeared to be black ice. This was placed in a container and over a few hours it melted and turned to what appeared to be bad smelling water. A grayish residue remained, which was saved and which was photographed. It resembled grey bird droppings and was crumbly and ugly looking.

The question is, was this an ice meteorite—possibly a fragment of a comet—the dirty ice of a comet fragment melting and evaporating, and leaving behind a residue has been reported elsewhere. Here is a photo of the residue left by the melting and evaporation of water, which may have composed an ice meteorite.

Rusty material from Arkansas (the rusty soil that continued on down): Northeastern Arkansas, except for Crowley's Ridge, is flat land—floodplains of a number of rivers that flow off of the Ozarks as well as the eastern part of the flood plain of the Mississippi River. Using the idea of Harvey H. Nininger, the author once made up notices offering monetary rewards for meteorites and distributed these through general stores in four or five counties of eastern Arkansas. The thinking was that in this flat, agricultural region lacking rocks, any rock found there would be noticed and just might turn out to be a meteorite. In the town of Swifton, on highway 67, I got a bite. A local high school teacher who also ran a rock shop informed me of an area where the soil was very rusty and might just enclose a hidden meteorite. Using a borrowed WWII mine detector, we did get a reading in the area described. We proceeded to dig with a pick ax in the area of the reading and soon encountered chunks of crumbly iron oxide (limonite). We continued to dig and the ocher became harder and more intense and also more difficult to dig into. Eventually we had to give up—the oxide becoming too hard and if it really was a buried meteorite it was really buried deep. Other activities intervened and I've often wondered if there really might be a large siderite buried in Jackson County north of Swifton, Arkansas.

Meteorites from Outside of the Solar System

One of the major constraints to the exploration of space is that of time. To travel even within the Solar System is currently difficult and costly, especially considering the energy requirements to overcome the earth's "gravity well." When interstellar space is considered, the element of time becomes seemingly insurmountable. With geologic materials (rocks if you like), however, time is of little consequence. **Time,** considered in terms of the necessity of vast amounts of it being required to travel interstellar distances, is readily available on a geologic scale. Long periods of time increase the possibility and probability of a space rock hitting another surface in some part of interstellar space. In other words, the landing of a meteorite from elsewhere in the Milky Way galaxy on the Earth is an improbable event that, with the passage of vast amounts of time, becomes probable. Of the latter, being an improbable event, the axiom of "given enough time an improbable event becomes probable" is cogent.

The concept of a meteorite originating from outside the Solar System, perhaps from somewhere else in the Milky Way galaxy, is generally discounted as a consequence of the low probability of this event actually happening, considering the vastness of interstellar space. There is a theoretical aspect to this concept that involves the passage of spans of geologic time and the introduction of megatime puts an interesting "spin" on the "equation." The possibility of the fall of an extra-solar meteorite taking place during one's life time is an exceedingly remote and improbably event, but **given enough time an improbable event can become probable**. Favored regions of the earth, such as Antarctica and dry deserts like the Sahara, accumulate and preserve meteorites that fall in these regions over spans of geologic time. Over most other parts of the earth, weathering and other atmospheric processes destroy meteorites shortly after they fall, especially the stony meteorites that usually don't last long at all. Atmosphere-free planetary bodies, like the moon, on the other hand, subject a rock to no weathering, so any meteorite that might land on such a surface without being pulverized would stay around. In contrast to the Earth however, the Moon provides no "cushion" to slow down an approaching object, so that on impact, impacting meteoroids produce a crater and in the process are usually pulverized or vaporized, unless they hit at a low, glancing angle where they might then survive. Other than this obstacle, the moon and other atmosphere bodies have been potentially accumulating space rocks for over four billion years. Under these circumstances, space rock accumulation (particularly extra-solar space rocks) over megatime becomes a possibility—becomes a possibility because an improbable event becomes a probable one with the passage of geologic time.

Meteorites as "Messengers" from the Earth and Megatime

Large impacts on the earth, like the impact that apparently produced the K/T extinction event, are capable of launching rocks into space. This is apparently the origin of tektites, those peculiar glassy space rocks that have become widely distributed in collections. Although it hasn't yet been proven, it is quite likely that Earth rocks reside on the Moon, possibly in considerable numbers as atmospheric processes that destroy Moon rocks on the Earth (Lunar Meteorites) don't exist on the Moon. Earth meteorites may also exist on Venus and Mercury. Their presence on Mars and the satellites of the outer planets is less likely, however, as for an Earth rock to reach these places, it would have to have sufficient speed (escape velocity) to overcome the gravitational pull of the Sun. Some earth rocks, however, ejected from our home planet probably did obtain such a velocity (especially with large impacts) and some probably impacted and pieces may now reside on these planets. With an escape velocity of 30 Km/sec or more a rock even has the potential of overcoming the Sun's gravitational field and leaving the Solar System entirely, especially with a gravity assist from Jupiter and Saturn. (The Viking spacecraft in 1997 and 1999 did just that). Once a space rock is in interstellar space, it **might** eventually impact on some distant planetary body in the Milky Way—its **hitting** such a "tiny" target being assisted by gravitational focusing. **Might** is, however, very speculative considering the vastness of interstellar space. However, again such an improbable event becomes more likely with the passage of megatime, considering that the increased probability of an improbable event, like interstellar matter exchange, **can become probable with the passage of geologic time.**

Snowball Earth

Rock strata some 650 million years old yield worldwide evidence of extensive glaciation. This is in the form of what are known as tillites, "fossil" glacial sediments. Other regions of the earth at this time show evidence of erosion having taken place on a grand scale. It has been hypothesized that some mechanism 650 million years ago took place that was capable of producing cold conditions that formed glaciers even near the equator and that large portions of the oceans may have froze solid. Various scenarios capable of producing such a massive "ice age" have been proposed, such as blocking of the Sun's energy with the encounter of a large dust cloud in the Solar Systems galactic orbit. Such a dust cloud, intervening between the Sun and the Earth, blotted out much of the Sun's energy, which otherwise would have been received by the Earth. This blocking effect generating extremely cold conditions. Hard evidence for such a scenario currently does not exist; however, continued looking in the right places (which might include other planets) might turn something up in the future to support this hypothesis.

This is the same ancient surface as shown above, but the weathered granite has been removed to a greater extent by weathering. Resistant-to-weathering rocks are embedded in the sandstone. These are rocks that laid on the surface of Missouri 550 million years ago. These rocks were embedded and preserved from weathering by being covered with sediment, which now is the 550 million year old sandstone in which they are still embedded. Such ancient surfaces, known in geology as an unconformable surface, may have accumulated meteorites besides these earth rocks, but only a very small portion of this surface is exposed (as it is in this road cut), so that the finding of one would be a serendipitous and lucky find.

Major Unconformity! This is a hiatus which separates almost a billion years of geologic time. The rock layer **below** the pronounced contact (line extending across the middle of the photograph) is granite formed (about) 1.4 billion years ago. This granite (and similar rocks) formed the surface of Missouri and other parts of the US Midwest 700 million years ago. On this surface, which existed for so long many things accumulated, meteorites are probably included.

Another view of this ancient surface of Missouri. The recessed rock is weathered granite. This rock is some 1.4 billion years old! Overlying it is sandstone, some 550 million years old, which has covered (was deposited upon) the granite and covered and "encapsulated" rocks which were lying on the surface of this granite (and on the surface of what is now Missouri), preserving them from weathering for 550 million years.

Close-up of the rocks that sat on the surface of Missouri 550 million years ago. Unfortunately no meteorites are to be seen in this small area (and if they were present, they now would only be a clump of iron oxide).

Another very ancient surface that accumulated rocks for millions of years. The rocky outcrop exposed along and by this Ozark stream is made up of rocks which accumulated on the earth's surface for millions of years; it is part of the same surface shown in the previous photos, but has accumulated many more rocks. Such an accumulation, if it were in a dry, arid environment should include meteorites. They would not, however, be found here as meteorites don't like water and oxygen. The Moon, Mars, Mercury, and the asteroids themselves have also accumulated rocks like this—they have no atmospheres (unlike the case in Missouri) and therefore weathering on them is nil. (A meteorite has actually been spotted in images transmitted from Mars.) On the earth, dry deserts like the Sahara and the frozen deserts of Antarctica provide the best places for preserving and collecting meteorites. This rocky outcrop along an Ozark river represents a possible earthly environment capable of accumulating (and preserving?) meteorites, but for the weathering. It is a situation that only would be recognized by a geologist. It is also an ancient surface associated with what is known as an unconformity.

This landscape was produced some 650-750 million years ago from 1.6 billion year old rocks. These rocks (or rocks like them) contributed cobbles to outcrops shown in previous photos. Some of this ancient landscape may have been carved by ice during the earth's "Snowball Earth" episode—a time when a profound ice age existed when even the oceans froze solid over large parts of the globe.

Left:
Snowball Earth? These rocks accumulated on an ancient rock surface in a low place. They may (in part) have been deposited by ice during the "Snowball Earth" ice age some 680 million years ago. The rocks in this accumulation were derived from nearby regions **prior** to 600 million years ago. The majority of the cobbles are from igneous rock about 1.5 billion years old. The overlying sedimentary rock, which covers this deposit of cobbles, is 550 million years old. These cobbles may have accumulated during any part (or all) of a 500 million year period of geologic time, the time period between the age of the overlying sedimentary rock and the age of the youngest cobble. The cobbles are all resistant-to-weathering local earth (Missouri) rocks and were it not for the earth's water and oxygen, meteorites would also accumulate on such a surface as the passage of a long span of time on a planetary surface increases the probability of an improbable event, and the landing of a meteorite in any small area is an improbable event. (Could iron-rich masses sometimes found with the cobbles be the oxidized remains of meteorites?) Such ancient surfaces on the earth are interesting and sometimes can show an accumulation of certain minerals over long spans of geologic time. A similar unconformitable surface in the Black Hills of South Dakota accumulated gold from the weathering of gold bearing rock over millions of years, the gold being concentrated by weathering of underlying gold bearing rock with gold accumulating on this surface as weathering progressed over a large span of geologic time. This gold became one of the sites of the 1876 Black Hills gold rush—and later became, by mining the gold bearing slate of which this ancient surface is composed, the Homestake Gold Mine near Lead, South Dakota.

Positron and Electron Tracks in a Cloud Chamber

This book is primarily about meteorites; however its just too tempting not to include subatomic particles, which, like meteorites, also are forms of matter (at least in some ways). These are particles that have been found to occur in cosmic rays, subatomic particles that originate from other parts of the universe.

When you watch a (cathode ray) TV screen or computer monitor, you are looking at an image "painted" by a stream of negatively charged particles (electrons). Electrons form the outer portion of all atoms—they are a fundamental part of matter as we know it. A cloud chamber is a device by which you can, in a way, actually see subatomic particles; that is, with a cloud chamber one can see the tracks made by moving subatomic particles like electrons and protons, as well as other subatomic particles.

The track on the first cloud chamber photo was made by an electron being deflected to the right, toward a positive charge, as electrons are negatively charged—being negatively charged the electrons will be attracted to a positive charge as unlike charges attract. The track on the second photo is just like that made by an electron, but it is deflected to the left, away from the positive charge. This is a track of a **positron**, a subatomic particle like an electron in all ways except that it has a **positive** charge rather than a negative one. It has been hypothesized that parts of the universe beyond the Milky Way galaxy might be composed of what is known as antimatter—matter made up of subatomic particles having opposite charges to that of our matter, matter which is made up of negatively charged antiprotons and antineutrons in the atom's nucleus and this surrounded by shells of positrons in place of electrons. Such antimatter, if it were to encounter matter like that in our Solar System, would convert both the matter and antimatter to energy and as a consequence, vast amounts of energy would be released in an immense nuclear explosion. (An anti-matter-matter reaction is the mechanism used for propelling the starship Enterprise in "Star Trek" episodes).

In 1908, a large explosion, known as the Tunguska event, did take place in the atmosphere above the Stony Tunguska River in Siberia. It knocked down trees over a 90 mile radius from "ground zero," yet no crater or other evidence of impact was found, the explosion apparently having taken place in the atmosphere. Some astrophysicists have hypothesized that this explosion may have been from a tiny speck of antimatter that made it to the earth from some distant portion of the universe. An antimatter-matter reaction would be a nuclear reaction, that is, one which would produce radioactivity and there is some evidence for such radioactivity in a carbon-14 spike supposedly found in trees that were growing in 1909. (The next such carbon-14 spike is found in tree rings from 1946). The Tunguska Event of 1908, any way you look at it, was a weird event.

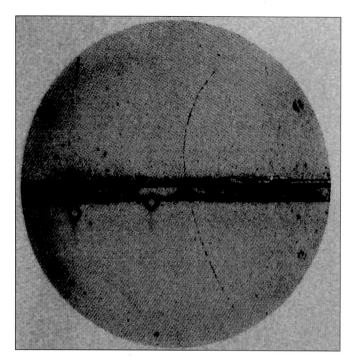

Cloud chamber photo showing tracks made by an electron. The dark, horizontal bar is a thin sheet of lead that the electron travels right through. The electron is deflected to the right as that side of the cloud chamber has a positive charge, which attracts the negatively charged electron.

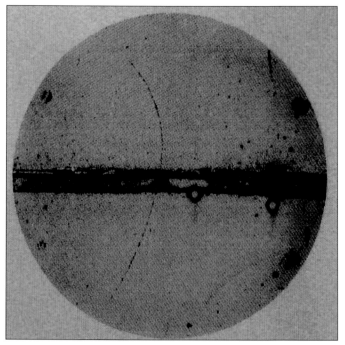

This photo was made by a cosmic ray particle identical to an electron but with a positive charge. This causes the particle to move to the left, deflecting it away from the positive charge (as like charges repel). Such particles, which originate from distant parts of the universe, could compose antimatter. Note that this particle, a positron, like the electron shown above, has also traveled through the thin lead sheet.

Trees blasted down from the 1908 Tunguska event in Siberia. The immense explosion, which took place June 30, 1908, apparently took place in the atmosphere; no evidence of an object impacting the earth's surface has been found. Atmospheric phenomena suggestive of a nuclear explosion accompanied or followed the explosion. It has been hypothesized that this event may have been caused by a very small speck of antimatter (matter composed of positrons, antiprotons, and antineutrons), which came from somewhere else in the universe and entered the earth's atmosphere, reacting with its matter and an equal amount of that of the atmosphere to produce a matter destroying nuclear reaction.

This is a rather "way out" explanation for what has also been explained as originating from a comet composed primarily of ice exploding upon its encounter with the earth's atmosphere. Occam's Razor would suggest this interpretation as most reasonable; however, J.B.S. Haldane's statement "The universe is not only stranger than we imagine, but is also stranger than we can imagine" may be applicable.

Lunar Meteorites

Lunar meteorites are currently important in working out lunar geology by the fact that they come from parts of the moon that have **not** been sampled by spacecraft, either manned or unmanned. They represent lunar material that was given an escape velocity sufficient enough to exit the lunar gravitational field and go into orbit around the earth. Lunar meteorites have come from all parts of the moon, including its far side—a part which otherwise has not been sampled by spacecraft. Lunar meteorites found to date have shown that samples brought back by the US and Soviet space programs were biased samples, especially with regard to the KREEP minerals (those containing potassium (K), rare earth elements (REE) and phosphorous (P)) which were present in amounts higher than normal in those samples brought back by spacecraft.

This rock essentially is a lunar basalt. Unlike terrestrial basalts, it is highly depleted in chemically combined water as well as in sodium and potassium—both elements being present in higher amounts in the makeup of earthly basalts. Lunar basalts are also higher in nickel, cobalt, tungsten, iridium, and phosphorous than are terrestrial basalts; they more closely resemble the composition of the earth's mantle—a characteristic that suggests that the Moon may have split off from the Earth some 4.5 billion years ago.

Lunar meteorite: This is a small slab from one of the unique meteorites originating from that bonanza of meteorite finds that came from the Sahara Desert of northwest Africa. Through trace element analysis and composition, these meteorites, known as lunars, have been found to have come from the moon. The process of spallation has been found to be responsible for samples ejected from the Moon and Mars (SNC meteorites of Martian origin). In some ways lunar meteorites might appear to be one of the more common types to be found; after all, you can look up and see this source with no trouble—its only 230,000 miles from the earth. Distance from the earth, however, is not of much consideration when it comes to meteorites. Time is a more important consideration and time to a space rock, or to any rock for that matter, is of little consequence! It takes millions of years of orbiting the sun for an asteroid fragment to acquire (by chance) an orbit which allows it to collide with the earth and result in a meteorite. Billions of asteroid fragments are out there, only a few rock specimens by contrast get blasted off from the Moon's surface and make it to Earth in a million years and those that do weather away on its surface, considering our water and atmosphere. Rarely will one of these lunars be collected as this one was. This lunar meteorite is a basalt porphyry, probably derived from lunar Maria-basalt plains, which were formed from mafic magma extruded from the moon's once molten interior—extrusion of lava from the maria being triggered by the impact of asteroids some four billion years ago.

References

Campbell, Carl E., Francisca E. Oboh-Ikuenobe, and Tambra L. Eifert, 2005. "Is the 'Paleocene' Clayton Formation in the Southeastern Missouri portion of the Mississippi Embayment the K/T Boundary Megatsunami Deposit?" (abstract) in Evans, Kevin R. et al., *The Sedimentary Record of Meteorite Impacts.* SEPM Research Conference at Southwest Missouri State University, Springfield Mo.

Evans, Kevin, et. al., 2005. "The Weaubleau-Osceola Structure and Weaubleau Breccia—Compiling evidence of a marine impact in the Sedimentary Record of Meteorite Impacts." Abstracts, SEPM Research Conference at Southwest Missouri State University. Springfield Missouri.

Haag, Robert A., 2003. *The Robert Haag Collection of Meteorites.* Self Published by Robert Haag Meteorites, P. O. Box 27527, Tucson, AZ 85726.

Melosh, H. Jay, 2001. "Exchange of Meteoritic Material between Stellar Systems." *Lunar and Planetary Science XXXII.*

Melosh, H. J., 2002. "Exchange of Meteorites (and life?) between Stellar Systems." Proceedings of the Rubey Symposium on Astrobiology.

Melosh, H. J. and W. B. Tonks (1994). "Swapping Rocks: Ejection and Exchange of surface material among terrestrial planets." *Meteoritics,* Vol. 28, p. 398.

Stinchcomb, B. L., 2005. "Knob Forming distinctive chert-breccia boulders of the Central Ozarks" (Abstract) in *The Sedimentary Record of Meteorite Impacts,* SEPM Research Conference, Southwestern Missouri State University, Springfield Mo.

Glossary

Directed Panspermia: Panspermia is the idea of the spread of a primitive form of life through interstellar space by its being incorporated in the inside of a porous meteorite. Directed Panspermia suggests a purposeful spread of such life by either an extraterrestrial civilization or a supreme being.

Gravitational Focusing: This is attraction by the gravity of a star or its orbiting planets (especially Jupiter-like planets) to a space rock that is moving toward the vicinity of the star. Gravitational focusing increases the possibility of an interstellar meteorite hitting an extremely small target like a planetary surface.

Lithopanspermia: Panspermia in which a microbe is protected from the hostile environment of deep space and its lethal radiation by being enveloped or included within the material of a meteorite.

Panspermia: The hypothetical idea that life on Earth was "seeded" by interstellar meteorites harboring primitive organisms, like microbes. This concept was first introduced in the 1960s by scientists Francis Crick and Stanley Orguel.

Spallation: The ejection of planetary material by high velocity extraterrestrial impact to become potential planetary meteorites. The process of spallation has taken place both on the Moon and Mars and is responsible for both Lunar and Martian (SNC) meteorites.

Unconformity and Unconformitable Surface: A break or hiatus in a sequence of sedimentary strata. An unconformity includes the existence of an ancient surface of the Earth which existed over some extended period of geologic time—a period of geologic time often being long enough to erode away whole mountain ranges. During this time span, meteorites of some sort might accumulate on such a surface. Usually material found on such an ancient surface consists of resistant-to-weathering rocks like quartz, but it also can concentrate resistant and insoluble material like gold and diamonds. Meteorites are items that might possibly accumulate on such a surface, especially in a dry environment and when this surface becomes covered with sediment, it may then preserve what has been buried. This, however, has not been a source of any meteorite to date in part as large, fresh areas of such surfaces normally are not exposed and large areas of ancient surfaces, like that of the Canadian Shield, would have had their meteorites weathered away long ago.

References

Bevan, Alex and John de Laeter. 2002. *Meteorites, A Journey through Space and Time.* Smithsonian Institution Press, Washington D. C. and London. ISBN 1-58834-021-X.

Carr, Michael H., R. Stephen Saunders, Robert G. Strom, Don E. Williams, 1984. *The Geology of the Terrestrial Planets.* SP-469. National Aeronautics and Space Administration (NASA), Scientific and Technical Information Branch. Washington D. C.

Cassidy, William A., 2003. *Meteorites, Ice and Antarctica—A Personal Account.* Cambridge University Press, Cambridge U K. ISBN 0-521-25872-3.

Cooper, Henry S. F. Jr. 1970. *Moon Rocks.* The Dial Press, New York.

Croswell, Ken, 1995. *The Alchemy of the Heavens-Searching for Meaning in the Milky Way.* Anchor Books, Doubleday Dell Publishing Group, Inc. 1540 Broadway, New York, N Y. ISBN 0-385-47214-5

Gallant, Roy A., 2002. *Meteorite Hunter, the Search for Siberian Meteorite Craters.* McGraw-Hill Co. New York-Chicago etc. ISBN 0-07-137224-5

Haag, Robert A., 1997. *Field Guide to Meteorites.* Self Published. Tucson, Arizona. ISBN 1053-8267. www.meteoriteman.com

Hutchison, Robert 1983. The Search for Our Beginning. British Museum (Natural History), London, Oxford University Press, Oxford England. ISBN 0-19-520435-5 OUP (USA).

_____ and Andrew Graham. *Meteorites.* Steerling Publishing CO., New York N. Y. ISBN 0-8069-0489-5.

Killgore, Kitty, 2002. *Southwest Meteorite Collection, a Pictorial Catalog.* Southwest Meteorite Press, Payson AZ. ISBN 0-9726921-0-X

Lauretta, Dante S. and Marvin Killgore, 2005. *A Color Atlas of Meteorites in Thin Section.* Golden Retriever Publications and Southwest Meteorite Press. ISBN 0-9720472-1-2.

McSween Jr., Harry Y.,1999. *Meteorites and their Parent Planets,* 2nd Edition. Cambridge University Press, Cambridge U K. ISBN 0-521-58303-9

Norton, O. Richard,1994. *Rocks from Space Meteorites and Meteorite Hunters.* Mountain Press Publishing Company, Missoula, Montana. ISBN 0-87842-302-8. **One of the most recommended and complete general works on meteorites, highly recommended!**

_____ and Lawrence A. Chitwood, 2008. *Field Guide to Meteors and Meteorites.* Patrick Moore's Practical Astronomy Series, Springer, London. ISBN 978-1-84800-156-5.

Regelman, Kenneth, 1995. *ARN's History of Meteorites.* Astronomical Research Network, 206 Bellwood Ave., Maplewood, MN 55117.

Reynolds, Mike D., 22001. *Falling Stars, A Guide to Meteors and Meteorites.* Stackpole Books, Mechanicsburg, PA. ISBN 0-8117-2755-6.

Wasson, John T., 1974. *Meteorites, Classification and Properties. (Minerals and Rocks).* Springer Verlag, New York-Heidelberg-Berlin. ISBN 30-387-06744-2.

Zanda, Brigitte and Monica Rotaru, 2001. *Meteorites, Their Impact an Science and History.* Cambridge University Press. Cambridge, United Kingdom.